Nandigram
Bio-cultural and Ecological Issues

Nandigram
Bio-cultural and Ecological Issues

B. N. Sarkar, Sumit Mukherjee,
Prabir Chowdhury, Shyamal Kumar Nandy,
Arup Ray and Nupur Sinha

Anthropological Survey of India
Government of India, Kolkata - 700016

GYAN

ISBN: 978-81-212-1259-5
Price ₹ 500/-
First Published, 2016

Published by
Gyan Publishing House
5, Main Ansari Road
Daryaganj, New Delhi-110002
Moble: 9811692060
Fax: 91-11-23285914
E-mail: books@gyanbooks.com

Printed at: G. Print Process, Delhi

Cataloging in Publication Data--DK

Courtesy: D.K. Agencies (P) Ltd. <docinfo@dkagencies.com>

Sarkar, B. N., author.

Nandigram: Bio-cultural and Ecological Issues / B.N. Sarkar [and five others].

pages cm

Includes bibliographical references (pages) and index.

ISBN 9788121212595

1. Peasants--India--Nandigram. 2. Eminent domain--Social aspects--India--Nandigram. 3. Anti-globalisation movement--Social aspects--India--Nandigram. 4. Human ecology--India--Nandigram. I. Anthropological Survey of India.issuing body. II. Title.

DDC 305.5633095414 23

Contents

Foreword

During the last couple of years the Anthropological Survey of India has expanded its horizon of study to cover various emerging issues in the country. The present volume is the outcome of one such study conducted by the survey based on assessment of social, economic and biological impact of the peoples' movement *vis-a-vis* violence of Nandigram-1, a Community Development Block in Purba Medinipur, West Bengal. Nandigram is a very old rural settlement in the deltaic South Bengal under the district of Purba Medinipur having a glorious past as evidenced from the inscription on *Tamralipta* (The ancient port city). In 2007, the proposal for land acquisition at Nandigram Block-1 to set up a Special Economic Zone (SEZ) caused social unrest resulting into a power struggle between the villagers and the local administration. Since then more than five years have passed, but villagers are still suffering from the impact of that incident.

A multi-disciplinary team consisting of socio-cultural anthropologists, physical anthropologists, human ecologist and psychologist from the Survey conducted this study using their interdisciplinary expertise and experience in understanding several emerging issues in Nandigram. An effort has been made in this study to understand and analyse the socio-economic and ecological background of the movement and counter-movement, and the after-effects of violence on people by evaluating their demographic, health and nutritional profiles, patterns of morbidity and mortality, psycho-social stress and disorder. The researchers have come out with very rich empirical data through detailed interviews and analysed the same in a systematic manner. The volume contains an outline of the emerging socio-economic, biological and ecological issues in the post-violence Nandigram which show worldwide concern for the amelioration of the local people.

The book has an important contribution to the study of contemporary social issues. I am confident that the publication will stimulate further discussion on the various important issues concerning to the land acquisition in our country, when we are trying for a new legislation on this. I am sure this volume would be of immense use to the social scientists, economists, welfare workers and administrators.

18th April, 2013
Kolkata

Professor K. K. Misra
Director
Anthropological Survey of India

Preface

The onset of globalisation and rapid development process has resulted in acquisitions of huge land where several cases of forceful evictions and involuntary surrender of lands occurred; and the people started struggling against the land acquisition. In recent years there has been accentuation of violent movements on such issues from people of varied ethnic, economic, socio-cultural and political affiliations across the country posing a slowdown in setting up of large development initiatives by public and private sectors.

A violent movement took place in certain villages of Nandigram Block-I after a controversial notification in connection with land acquisition of 28 villages were issued to set up a Special Economic Zone for petro-chemical hub in that area. It had resulted in a clash between two groups and the villagers suffered physically, mentally and lost their lives in that incident. It had happened nearly five years back, i.e., in 2007, but villagers are still suffering from that incident. It is rather amazing to note that in absence of any official notification directly from the relevant department of the State Government an unsuspected peoples' movement flared up. This signifies the immense sensitivity of the issue of involuntary displacement and the mistrust on the system of land acquisition by the government authority in the present day.

Nandigram is a very old rural settlement in the deltaic South Bengal under the district of Purba Medinipur having a glorious past as evidenced from the inscription of *Tamralipta* (present Tamluk town) and mainly in the descriptions on the visits of Emperor Asoka (Circa 261 BC) and famous Chinese traveller Fa Hien (640 AD). Nandigram has a long history of struggle and rebellion starting from 'Chuar Bidroha' or the revolt of the *paiks* (country-soldiers) in 1799 AD; active freedom struggle against the British Rule in

India; agricultural share-croppers (*bargadars, bhag chashis, adhyars*) launched a movement popularly known as Tebhaga Andolan in 1946-47. The area has a natural strategic geographic location for business, farming and industry. The Hugli (Bhagirathi) and Haldi Rivers cover the eastern and northern edges of Nandigram, and thereby the land is fertile by both the rivers. The low-lying marshy land provides attraction and facilities to some agrarian and other fishermen communities who came here from the south by the land routes as also the *Feringhees* or the Portuguese pirates who came by ships and plundered this land from time to time. The entire Nandigram economy mainly evolves around agriculture, beetle leaf cultivation, fisheries and brick manufacture.

The present volume is an outcome of the research study carried out in Nandigram, in Purba Medinipur district of West Bengal. The project was undertaken on the patronage of Shri Jawahar Sirkar, the then Secretary, Ministry of Culture, Government of India and Professor K.K. Mishra, Director, Anthropological Survey of India (AnSI). The study was conceived with an idea of impact assessment of the people recently affected by the movement related primarily to land acquisition for development project initiatives.

A multi-disciplinary team constituted involving Cultural Anthropologists, Physical Anthropologists, Human Ecologist and Psychologist with an objective of comprehensive study. The chapterisation itself reflects an integrated systematic approach in conformity with the objectives and specific ground realities. Altogether 275 households were selected for this investigation from the three most affected villages Sonachura, Sonachura Jalpai and Garchakraberya.

The focal point of the entire study was on three major research queries: (i) the socio-economic and ecological background of the movement and counter-movement; (ii) the after effects on health and nutritional profile, patterns of morbidity and mortality, psycho-social stress and disorder of the villagers by the help of case studies; and (iii) the socio-political construction of the movement under the backdrop of conflicting situation due to proposed land acquisition at Nandigram. Therefore, the study was intended to look on the historicity of the problem not only to interpret the episode but to discover the root of the problem. On that backdrop the study will help the planners and policy-makers a lot in future apart from its academic interest. With the LARR Act, 2013 already in force since January 2013 and Anthropological Survey of India, has already submitted the Social Impact Assessment Module to the Government of India, this volume will certainly attract attention of the policy makers and development executives. On the other hand, for the people of Nandigram this document may come handy who are still in doubts and dilemma even after the apparent change in the outlook of the present West Bengal Government.

The spirit of the study was an academic endeavour to identify the impact areas and not merely a post-mortem of the incidence. This tragic outcome of Nandigram episode was unwarranted and uncalled for. The local people of Nandigram were the worst and direct sufferer in one hand. On the other, the state as well as the entire society is still suffering from trauma and indecisiveness where the land issue and industrial development is under stake.

The study reveals at least three major interest groups in Nandigram incidence. The first being the erstwhile landlords or *mahajans* who have lost their holding through The West Bengal Land Reform Act, 1955 and had negative or no interest on the riverside land given to *bargadars* or *pattadars*. Secondly, the rich cultivators, having larger quality land holdings and good returns thereof, can be seen as an aspirant group for highest possible compensation for land. Thirdly, the legal and unofficial *bargadars*, though suffering as marginalised holders, were found split in two groups. The registered one could visualise direct compensation whereas the other one found it as the last nail on their coffin.

At the onset of Nandigram unrest, the area of protest was limited to Sonachura Gram Panchayat to Kalicharanpur Gram Panchayat in anticipation of the land acquisition. After 14th March, 2007 (The incidence of police firing) the shift of interest groups were noticed. With the growing strength of the people opposing the project the then ruling political party began suffering from fear of losing political authority on the area.

The present study indicates that violence affected families have suffered a lot almost in every sphere of life, i.e., social humiliation, mental unrest, physical assault, economic disorganisation. The affected villagers are gradually overcoming the situation through adjustment with the local village society. Some demographic changes are discerned in the age-sex structure of the population. The anxiety and depression were the main psycho-social stress developed by the people in Nandigram area at that time.

It is evident that this incident cropped up due to lack of proper understanding of the socio-economic and socio-political processes in general and current land use pattern of that area in particular on the part of relevant development or administrative bodies. Land (*jami*) is very precious to them both for the land-owners and land-less peoples, viz., *bargadars*. Land-less people are indirectly depended on other's land for agricultural labour (*khet majur*) and contract cultivation (*chukti chas*). Land is to be seen not only as a commercial point of view but also social point of view and it has immense social significance. The present report has attempted to depict and assess the current resource base and land use in a spatial perspective generated through imagery-based mapping of the area in GIS platform. The demographic and economic dimensions of the human resource along with the current land use and cropping pattern at village level have thrown much light on the issue of land. This will help in assessing and understanding the updated value of land needed for an effective and reasonable resettlement and mitigation policy.

Who and what triggered the Nandigram violence is yet to be understood conclusively. But there are essentially two contrasting interest groups in this highly contentious issue having widely varied view points from one another. Each has its merits and lacunae as mentioned in the course of this report. The first scenario is the accounts of *intense* spontaneous protest and agitation over the issue of proposed land acquisition under Bhumi Uchchhed Pratoridh Committee (BUPC) or the Land Eviction Resistance Committee and the other is the top leadership of the then ruling party, determined to rule out all oppositions against industrialisation. However, the villagers of the Nandigram still believed that such a tragic violence could have been avoided with a mutual trust and proper mitigation plan. The movement itself is an account of the peoples' reaction to such acquisition drive irrespective of location and community involved in our country. Hence the present volume, instead of its limited scope, seems to extend a documented evidence of an important instance of such movement in the perspectives of their demographic history, social, economic and political dimensions.

B. N. Sarkar
Sumit Mukherjee
Prabir Chowdhury
Shyamal Kumar Nandy
Arup Ray
Nupur Sinha

Acknowledgements

The authors gratefully acknowledge the initiative and guidance extended by Shri Jawahar Sircar, the then Secretary of the Ministry of Culture to undertake this kind of study. We are also grateful to Professor K.K. Misra, the then Director Anthropological Survey of India for his guidance and support. This publication of this volume is made possible by the valuable inspiration and official support from the present Director Shri G.S. Rautela. Thanks are also due to Shri D.N. Pandey, Head of Office, Anthropological Survey of India (AnSI) for providing all logistic support for the study.

We gratefully acknowledge the immense help and cooperation received from the various offices of the District Administration particularly the Land and Land Revenue Department, Department of Health and Family Welfare, Block Development Office, Nandigram Block-I, Sonachura and Kalicharanpur Gram Panchayats for providing all secondary source of information and assistance whenever required during the field investigation. Thanks are also due to Abdur Sattar, Swasthya Karmadhakshya, Mujibar Rahaman, a local leader, Professor R.N. Maity former Savadipati and Professor Krishnendu Bhattacharya, Department of Anthropology, Nandigram College for their help and suggestion during field investigation at Nandigram. Our thanks also due to Smt. Subhra Bhattacharya, Shri Tapas Biswas and Smt. Arpita Jana for their help and co-operation during the health survey at Nandigram and Dr N. K. Duary and

Dr Umesh Kumar of the AnSI for their suggestion during the preparation of this report. Last but not the least we are extremely thankful to the informants both women and men who spent their valuable time and responded with tremendous patience during our field investigation.

B.N. Sarkar
Sumit Mukherjee
Prabir Chowdhury
Shyamal Kumar Nandy
Arup Ray
Nupur Sinha

List of Tables

List of Figures

1

Introduction

The industrialisation and urbanisation processes are almost common in all the countries of Asia, Africa and Latin America during their colonial and post-colonial periods of history. These two major processes uprooted peasants as well as artisan communities from their natural resource base like land, water and forest (Guha, 2007). In rural India, the agriculture still continues to be the major livelihood activity with a distinct cultural contour even after rapid industrialisation (Chakraborti, 1985). This situation is still relevant to the present context. The recent onset of globalisation has transformed the old ideas about socio-economic conditions of the local people. In February 2005, the Indian Parliament passed the Special Economic Zones (SEZ) Act. A lot of land is being acquired by the governments to keep the land at the disposal of the developers of the SEZs. But, proper rehabilitation and resettlement (R&R) schemes have not been prepared in each case and particularly the people being affected have not been taken into confidence. Therefore, the forceful evictions and involuntary surrender of lands occurred; and the people started struggling against the land acquisition.

It is obvious that any act of dispossession from one's own means of production results in not merely the loss of land but importantly economic security, social status, empowerment achieved through earlier movements and land reforms, home and kinship.

According to Cernea (1999), "Expropriation of land, removes the main foundation upon which people's productive systems, commercial activities and livelihood are constructed. This is the principal form of decapitalisation and pauperisation for most rural and many urban displacements, who lose this way both natural and manmade capital".

Particularly, the SEZs in India have recently turned out to be a landmine as is evident from the Salim's proposed chemical hub project at Nandigram or Tata's project at Singur in West Bengal in recent times, it draws everybody's attention. Nandigram[1] is famous in the whole world for vibrant people's agony, which was burst out by repeated violence on the issue of proposed land acquisition. A power struggle took place at Nandigram Block-I in connection with land-related issues to set up a special economic zone in that area. It has resulted in a clash between two groups and the villagers suffered physically, mentally and lost their lives in that incident. It had happened nearly five years back, i.e., in 2007 but villagers are still suffering from that incident.

1.1 Land and People

Geographical Aspects of the Region

Geographically, Purba Medinipur alluvial plain is defined as a part of mature delta forming the south-western end of the Bengal delta. The region has a south-easterly slope. The river Keleghai and Kasai, originated from Medinipur upland on the west, join together to form the river Haldi which flows as a large tidal channel into the river Hugli near Nandigram. A portion of Odisha coastal canal runs through the area. Due to tidal and maritime impact, the soils here are mostly saline and range from silt to clay. The soils are grouped in two major types, i.e., (i) *Jal-jami* which includes both low and medium lands utilised largely for cultivation of all crops; and (ii) *Kala* or highland which is mostly the homestead and adjacent areas holding the betel leaf vines and different fruit orchards. At present there is hardly any trace of mangrove vegetation or any kind of natural vegetation in and around Nandigram. Oppressive hot summer, high relative humidity and well-distributed rainfall during the monsoon season are some of the characteristics of the climate of Medinipur.

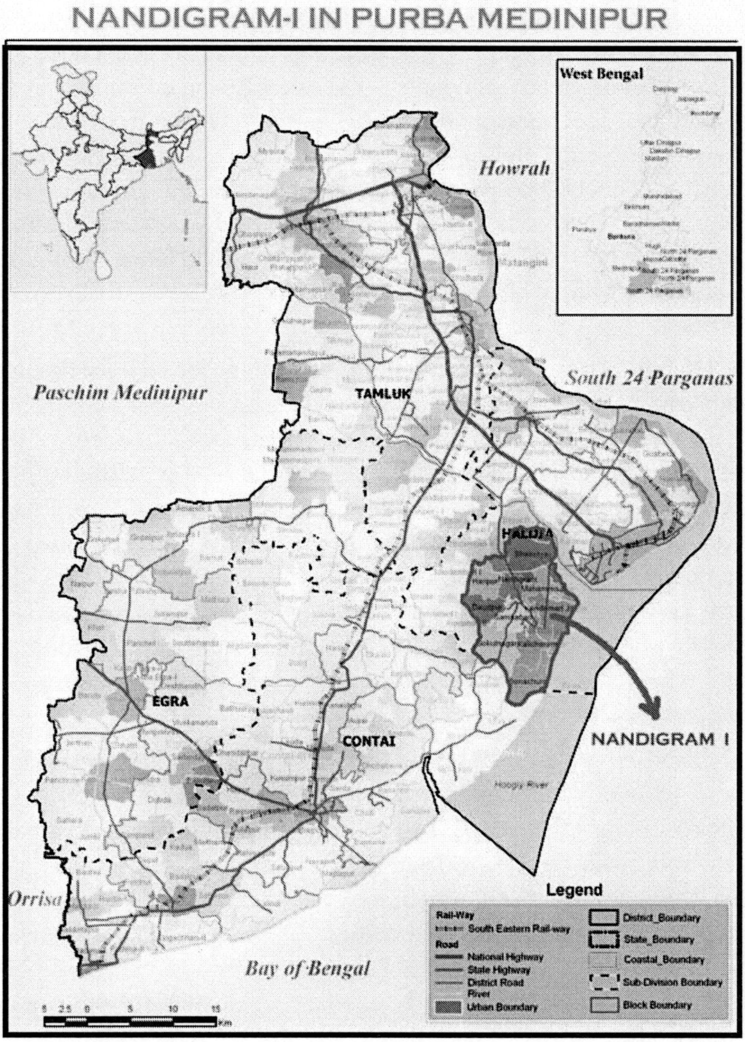

Fig. 1.1. Location of Study Area.

The seasons are well marked. The cold weather begins from about the middle of November and lasts up to the end of February. The summer season lasts from March to May. June to September is the south-west monsoon season. October and the first half of November constitutes the post monsoon season. At the beginning and termination of the rainy season, the climate is oppressive and the weather is cloudy and sultry. The average annual rainfall in the district is 1525.5 mm. The annual rainfall varies from place to place. The rainfall during the monsoon months June to September constitutes about 74.0 per cent of the annual rainfall. On an average there are 73 rainy days in a year. The area is exposed to strong monsoon winds and is visited by devastating cyclones throughout its history with a huge effect of *Aila* in 2009 still fresh in the memory.

Nandigram has a natural strategic geographic location for business and farming. The Hugli (Bhagirathi) and Haldi, which is a confluence of the river Kasai or Kansabati and Keleghai, cover the northern and eastern edges of Nandigram, and thereby the land is fertile by both the rivers (Fig. 1.2). Most of these rivers are also navigable and especially the estuary of the river Hugli offers greater facilities for navigation by boats, merchant vessels and ships, which provide easy communication with neighbouring districts across and with people of distant lands by sea. This migration route encourages new social and economic relations between the local and various other groups of emigrants.

The low-lying marshy land provided attraction and facilities to some agrarian communities who came here from near south by the land routes and other people like fishing communities and the Paundra Kshatriya, who came here from the district of 24-Parganas, as also the *Feringhees* or the Portuguese pirates who came by ships and plundered this land from time to time. Portuguese pirates attacked the area during 1610 to 1673. L.S.S. O'Malley has written, "The Portuguese had occupied the Hijli island, and selling forth in their galleys and ships, in combination with the Arakanese pirates, committed serious depredations on the coast, raiding the villages and carrying away men and women to sale them as slaves" (O'Malley, 1913). Thus, the external aggression had developed a hard psyche of the people to counter any difficult situation.

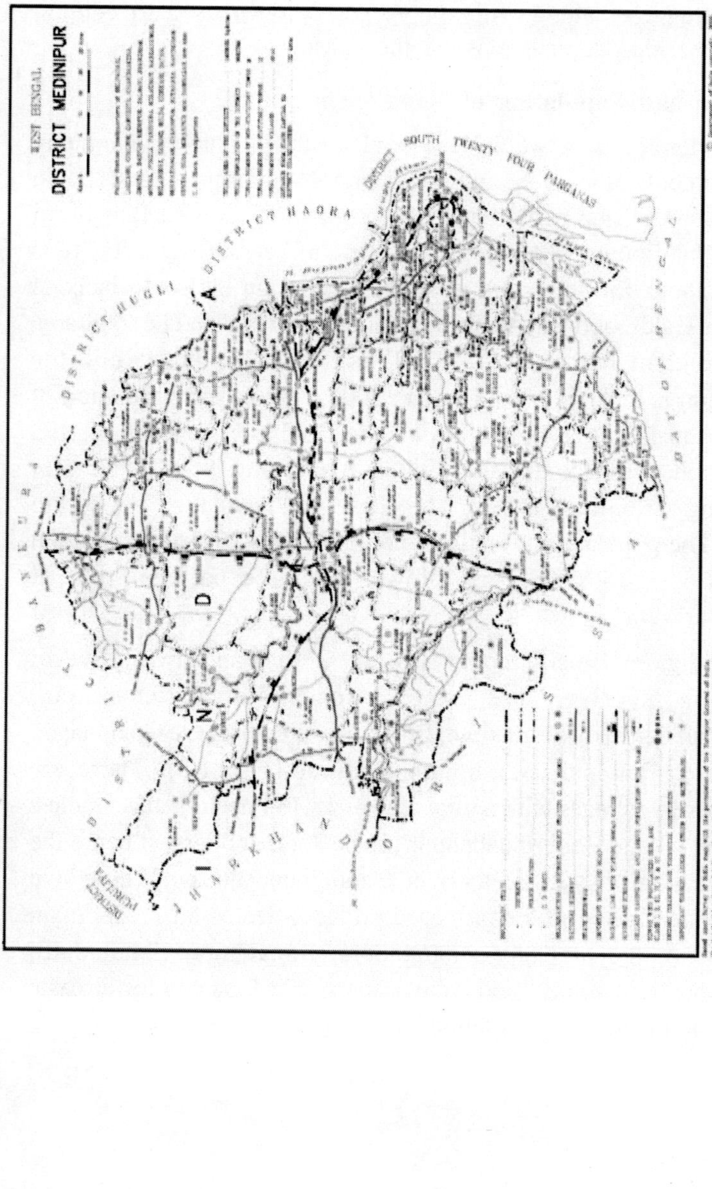

Fig. 1.2: Map of Undivided Medinipur District Showing Major Rivers, Transport Network, Major Settlement etc.

The agrarian communities first came to the low-lying swampy region and had taken to manufacture of salt which was a profitable trade of the local *zamindars*, the British as well as the ancient Mohammedan and Marhatta rulers. The coastal region particularly Hijli and Khejuri, adjoining to the Nandigram, became trading places of the Portuguese, who had their settlements and churches there. The Portuguese, along with the Arakan pirates used to commit robbery serious depredations on the coast.

1.2 Area and Population of Nandigram

Nandigram, as a whole is a rural area with three Community Development Blocks, Nandigram-I and Nandigram-II and III in Haldia sub-division of Purba Medinipur district of West Bengal, of which Nandigram Block I is the most affected (Fig. 1.1). It is located about 150 km south-west of Kolkata, on the south bank of the river Haldi, opposite the industrial city of Haldia. The total area of Nandigram (Block I, II & III) is 413.74 sq km, while the population is 439,077. According to 2001 Census, the population of Nandigram-I was 174,691 which are 3.95 per cent of district population. This comprises 89,053 males and 85,638 females indicating a sex ratio as 962. The density of population per sq km is 961. The population growth rate during 1991–2001of Nandigram Block-I was 14.87 per cent lower than the national rate of 21.34 per cent.

Nandigram Block-I comprises ten Gram Panchayats; five of which on the river side namely: Sonachura, Kalicharanpur, Kendamari-Jalpai, Mahamadpur, Bhekutya and others are Gokulnagar, Samsabad, Daudpur, Nandigram, Haripur (Fig. 1.3). There are altogether 99 villages with a total area of 18,184 hectors, the smallest being the village Balibrindabanpur (11 hectors) and largest being the Nakchira Char (1,438 hectors) in Nandigram Block-I. These two villages are located in the proposed project affected area. The main villages in this area are Bhangabera, Sonachura, Saudhkhali, Maheshpur, Gokulnagar and Adhikaripara. Muslims and lower caste Hindus dominate the population.

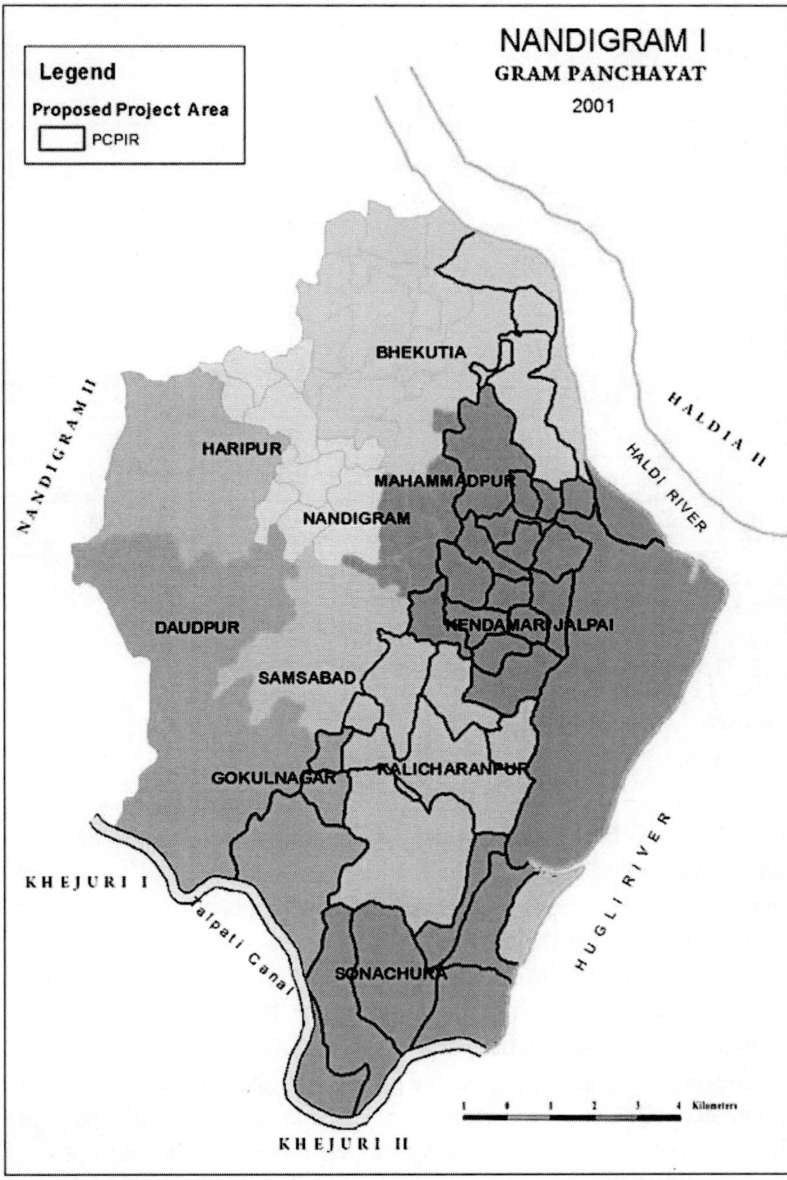

Fig. 1.3. Distribution of Gram Panchayat in Nandigram Block-I.

Table 1.1: Gram Panchayat Wise Population Distribution in Nandigram C.D. Block-I.

Sl. No.	*Gram Panchayat*	*Population**	*Percentage of Population to Block-I*
01.	Bhekutia GP-1	26,963	13.49
02.	Mahammadpur GP-2	18,820	9.42
03.	Kendemary GP-3	23,274	11.65
04.	Nandigram GP-4	18,077	9.04
05.	Haripur GP-5	15,639	7.83
06.	Gokulnagar GP-6	22,280	11.15
07.	Samsabad GP-7	18,650	9.33
08.	Daudpur GP-8	18,328	9.17
09.	Kalicharanpur GP-9	17,823	8.92
10.	Sonachura GP-10	20,003	10.00
	Total	**1,99,857**	100.00

Source: BDO, Nandigram-I

1.3 Demographic Profile of the Region

Density of Population

The Nandigram C.D. Block-I, with a density of population 961 per sq. km, ranked sixteenth in the Medinipur district as a whole in 2001. The village level density of population of the proposed SEZ area is an indicative of the habitation load and the land value and demand of the area, which is essential to assess the resettlement and rehabilitation scheme for the acquisition. The map (Fig. 1.4) shows that population density is relatively lower than the average, i.e., 1,039 per hact in the villages along the Hugli River in general. Nilpur (3846), Gopimohanpur (2132) and Garchakraberya (1924) in the north central part of the block are the top three densely populated villages. On the other hand, Nakchira Char (65) has the lowest density being a newly developed river shoal located at the southern bank of Haldi-Hugli confluence. Likewise, Sonachura Jalpai (458), Gangra Char (65), and Gangra (171) are among the lowest density villages located in south-eastern end of the area.

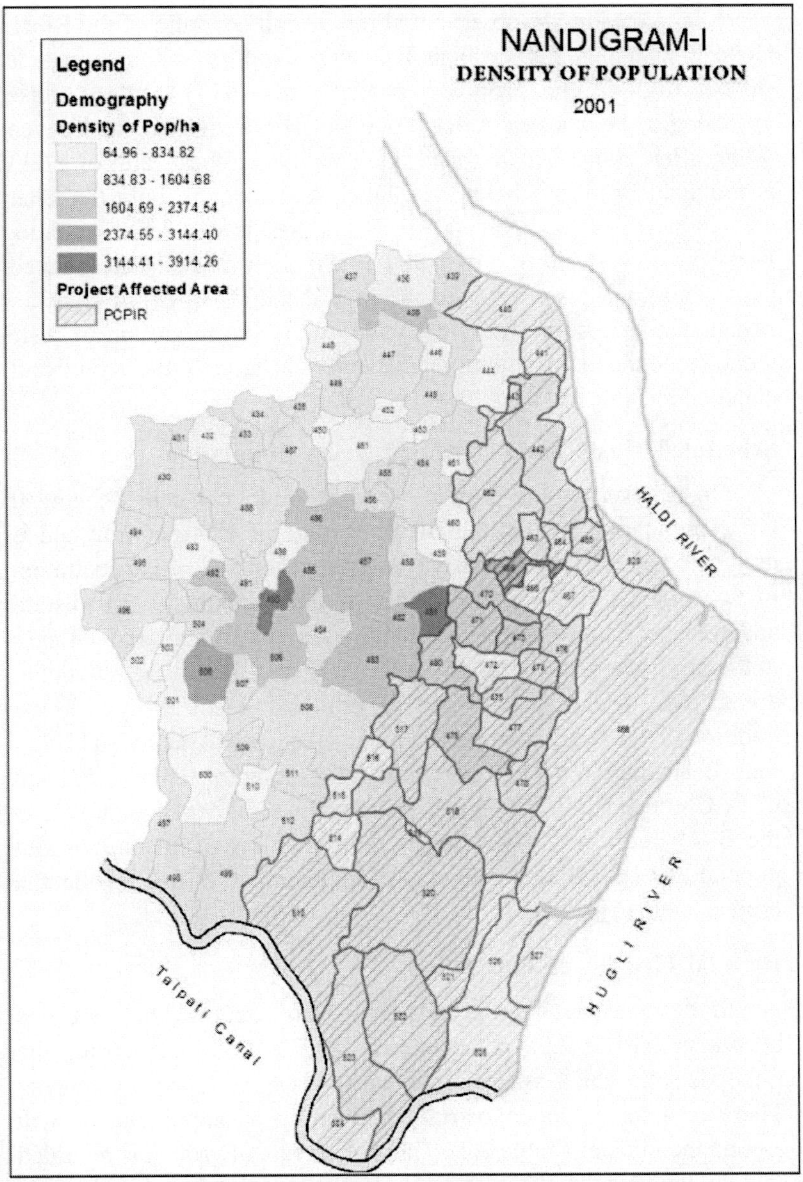

Fig. 1.4. Village-wise Distribution of Population Density

Sex Ratio

It is interesting to observe that the overall sex ratio of the Block is 962, which is higher than the state average and the ratio is further high in the proposed project area (973). Out of these 19 villages, bear a sex ratio above the Block average and seven villages had more female than male persons. Highest ratio of 1,333 is found in Bolibrindabanpur in the central part and only the Char area in north-eastern part of the Block has lowest level of ratio. Even the lowest sex ratio of 909 in Subhani Chak is an isolated case. This higher sex ratio in the area demands in depth study to reveal whether the non-resident male due to out-migration of male workers, which is a common phenomenon here, is the significant cause.

Scheduled Caste and Scheduled Tribe Population

On an average the project area has 21.45 per cent Scheduled Caste population against the Block average of 19.4 per cent and is dominated by Paundra Khsatriya and Namasudra. Clear concentration of Scheduled Caste population is over the southern, south-western and western parts of the Block with a few more scattered villages in the north-western corner of the Block (Fig. 1.5a). In the remaining areas, S.C. are distinctly sparse coinciding roughly with the traditionally Muslim dominated areas like Garchkraberya (11.98), Jadu Bari Chak (0.0), Usman Chak (0.0), Gopimohanpur (0.27) and Chak Chilinga (0.0). Scheduled Tribe population is almost absent in the Block with only two villages returning less than one per cent population, i.e., Gokulnagar and Kalicharanpur falling inside the project area (Fig. 1.5b).

Decadal Growth of Population

In these 37 villages proposed for land acquisition there has been a growth of 8.00 per cent from 73,528 to 79,413 during the same decade which was marred with severe violence and unrest. There is a lot of spatial variation in terms of population growth among the villages (Table 1.2). The highest rate of growth is recorded in two villages i.e., Baranagar (126.63) and Keshabpur (102.36) located at the centre of the Block followed by Char Kendemari (52.02), Sonachura Jalpai (44.86), Jadubari Chak (44.55) and

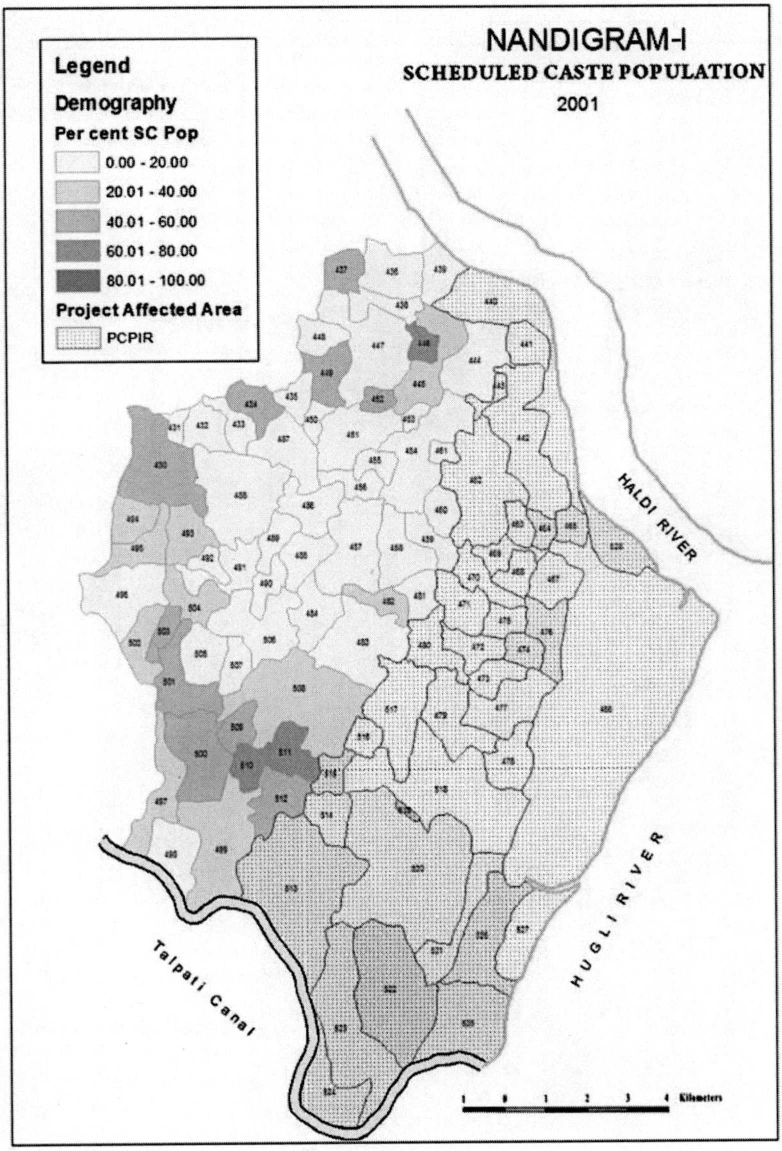

(a)

Fig. 1.5a Distribution of Schedule Caste Population

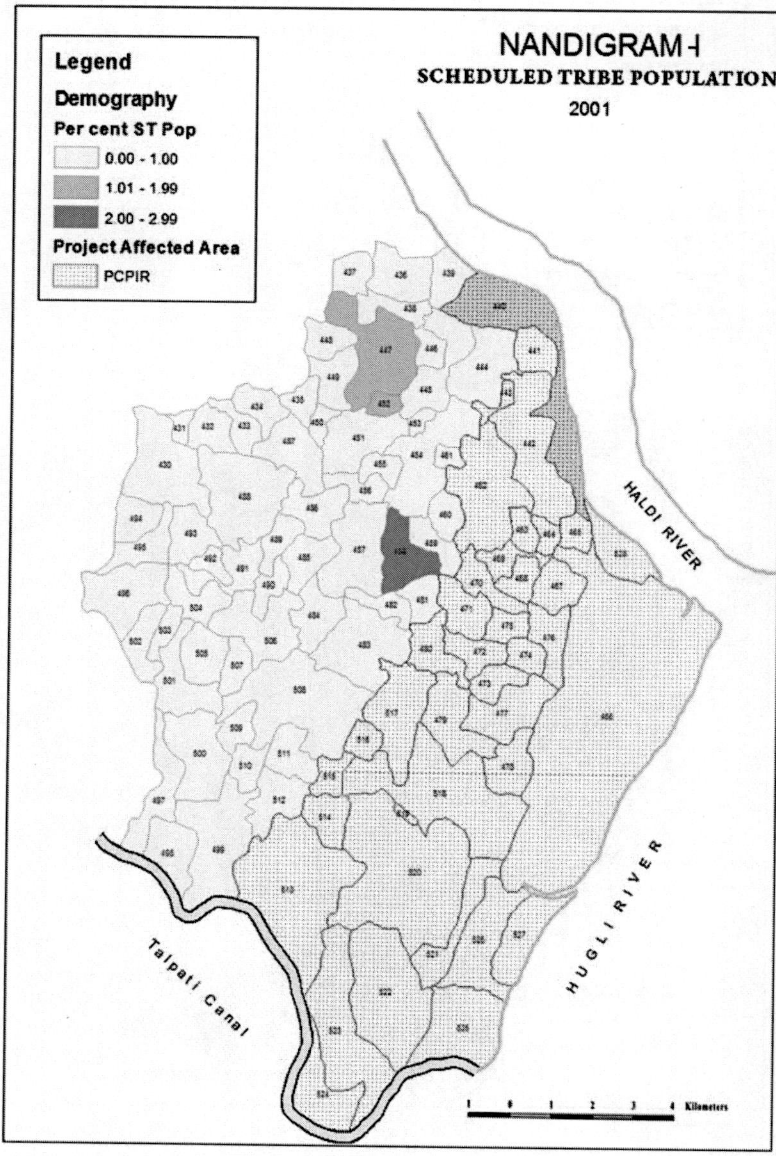

(b)

Fig. 1.5b: Distribution of Schedule Tribe Population

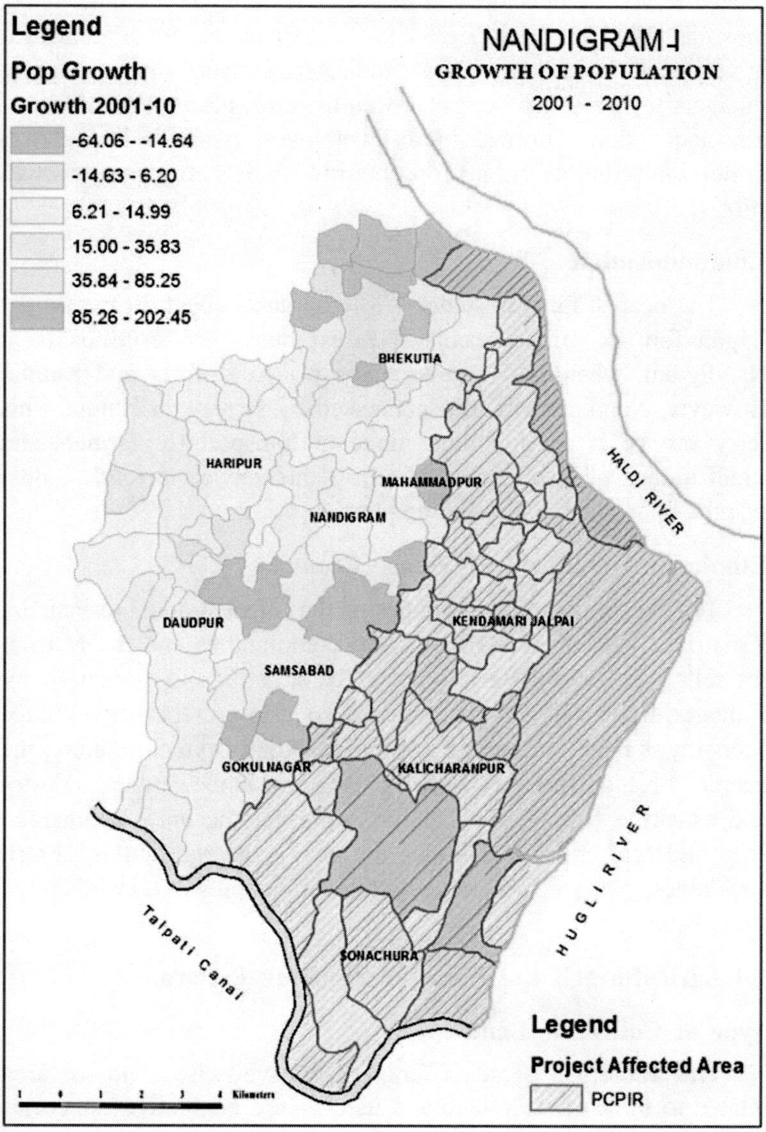

Fig. 1.6. Growth of Population (2001–2010).

Sonachura (31.89). It is interesting to note that the southern Char areas are showing a higher population growth which is most likely related to speedy development of prawn fishery. On the other hand, most disturbed areas like Keshabpur and Baranagar the high growth is very significant and probably indicate the return of non-resident villagers to strengthen their claim on the birth place in suspicion of any acquisition. Further, micro-level investigation is required to understand the driving force behind such pattern of growth (Fig. 1.6).

Communication

The nearest railway station—Mograjpur is about 23 km on the Digha-Tamluk railway route. Nearest busy bus terminus is at Nandigram, wherefrom regular buses ply to Kolkata and Kanthi. However, Nandigram is also connected by ferry with Haldia. This ferry service is an important mode of transport for farmers and small traders of Nandigram, who uses this service to reach Haldia market for selling their commodities.

Ethnic Composition and Village Economy

The predominant communities of the affected area are Paundra Kshatriya, Mahishya as Hindu caste communities and the Muslim (as religious group). Besides, other communities like Namasudra, Kaibarta, Rajbanshi, Dhopa, Mahato, etc. also present there. Village economy is mainly based on agriculture. Apart from agriculture, the people of Nandigram are engaged as labourers in the garment industry and estuarine fishing. Betel leaves represent the only commercial crop and brick kilns constitutes the only industrial activity. Fresh vegetables, rice and fish are being regularly supplied to Haldia town from Nandigram.

1.4 Agricultural Land Use and Cropping Pattern

Type of Cultivable Land

The Table 1.3 presents Gram Panchayat-wise share of area related to the cultivable land and its different qualitative categories in the Nandigram Block-I. Out of ten GPs in the Block, the five river side panchayats are in question, namely, Bhekutia, Mahammadpur, Kendamari Jalpai, Kalicharanpur and Sonachura share

Table 1.2: Population Growth and Economic Pursuits of Proposed Project Affected Villages

Village ID	*Village Name*	*Population 2010*	*Population Growth 2001-2010*	*Total Work Participation Rate 2001*	*Per cent Cultivator 2001*	*Per cent Agricultural Labourer 2001*
441	Basuli Chak	838	9.31	24.77	8.95	51.05
442	Dinabandhupur	3523	9.37	28.13	30.79	44.92
443	Moujbrindabanpur	568	9.09	28.21	42.18	31.29
463	Madhabpur	823	15.24	21.71	8.39	54.84
464	Parbatipur	161	13.10	29.58	2.38	64.29
465	NayaChak	915	15.91	32.95	23.85	58.46
466	Kenda Mari Jalpai	6276	-2.08	25.89	14.77	50.27
467	Baincha Bari	750	15.00	19.79	6.98	60.47
468	Hazra Kata	495	14.66	23.38	1.98	35.64
469	Nilpur	2036	15.07	23.40	4.11	41.30
470	Hosenpur	1999	15.69	20.78	1.95	51.53
471	Gopimohanpur	2988	13.82	24.57	7.75	27.13
472	Bhasur Kata	838	28.20	24.31	1.26	91.19
473	Subhanichak	957	25.96	29.08	19.91	26.70
474	Mirza Chak	912	25.86	25.79	9.09	50.27
475	Raja Ram Chak	1110	9.64	23.62	5.86	76.15
476	Shyam Sundari Chak	1147	9.88	36.21	43.92	35.98
477	Usman Chak	2478	6.53	25.06	27.96	40.65
478	Jadu Bari Chak	1696	3.71	34.37	12.28	32.38
479	Gar Chakra Berya	3881	4.48	26.41	11.11	57.59
480	Chilinga	2399	14.00	24.48	6.60	54.56
513	Gokul Nagar	7087	12.53	30.66	24.39	36.04
514	Keshabpur	942	31.89	27.45	59.18	23.98
515	Bara Nagar	719	44.86	27.62	87.59	2.19
516	Chak Chilinga	285	18.62	31.67	2.63	52.63
517	Kanchan Nagar	3618	13.98	24.20	10.81	30.08
518	Jalpai Part-VII	6696	9.52	29.34	19.40	42.31
519	Balibrindabanpur	23	9.25	28.57	0.00	100.00
520	Kalicharanpur	4494	-44.55	33.65	32.05	35.31
521	Saudikhali Jalpai	1312	11.27	38.51	2.86	59.25
522	Gangra	4523	16.03	33.20	33.62	46.60
523	SonaChura	6082	14.64	39.13	41.14	37.76
524	Sonachura Jalpai	1006	14.62	26.99	18.14	36.71
525	Gagra Char	1043	52.02	51.02	52.86	6.57
526	Saudkhali Char	1564	102.36	27.81	6.51	82.79
527	Nakchira Char	1033	10.65	29.98	8.21	54.29
528	Charkende Mari	2198	126.63	26.08	60.47	26.48
Total	**Nandigram - I**	**189691**	N.A.	**28.81**	**21.06**	**39.33**

Source: Primary Census Abstract 2001, Medinipur and BDO, Nandigram-I

62.70 per cent of the total Block area. Further, Kalicharanpur and Bhekutia are top two in size with a combined share of 32.65 per cent. On the other hand, excepting Bhekutia (82.10 per cent) all other river side GPs hold less than the Block average (70.38 per cent) of cultivable land. In terms of irrigated land, as well, these five GPs have a lower proportion than the average (15.74 per cent) with a clear exception of Mahammadpur (25.63 per cent) and Kendamari Jalpai (18.35 per cent). Hence, it can be deduced that the river side GPs proposed to be affected by the SEZ project has a marginally lower quality of cultivable land as per the traditional mode of agriculture recorded by the Agriculture Development Office.

This estuarine alluvial zone is further categorised as per various other characteristics of the land and soil. In terms of relative height, micro-topography and amount of moisture content the cultivable land is classified into three simple categories, viz., low, medium and high lands. Naturally, the low land category is more *moist* and seasonally water logged and even shows higher salinity on the river side areas. This category shares only 8.09 per cent of the total area and the share is further low in those five GPs (Fig. 1.7).

The major part (73.49 per cent) of the Block is under medium type land which is found most suitable for cultivation. Such medium land is present within a range of 63 to 77 per cent in those five GPs which is lower compared to the remaining inland GPs.

The high land category includes only 18.42 per cent of total land which, in reality, are the physically altered by human activities like building earthen embankments or raising homestead land with soils available while digging out ponds and reservoirs. Interestingly, this type of land is found at higher proportions in those five GPs compared to other five. This type of land is mostly utilised for betel vines, plantation, orchards and vegetable cultivations using the tank water.

The area is situated beside the confluence of Haldi and the Hugli Rivers. Creeks and canals crisscross the area which provide them sufficient water for cultivation. Other characteristic feature of the landscape is that the houses are surrounded by homestead lands (*kalabari)* which is used for vegetable cultivation and some ponds are also there. One household is distanced considerably from the other which is surrounded by cultivable land.

Table 1.3: Area and Type of Cultivable Land in Nandigram I (GP-wise) 2009-2010.

GP_Id	Gram Panchayat	Geographical Area (Ha)	Per cent cultivable land	Percent irrigated land	Per cent low land	Per cent medium land	Per cent high land
1	Bhekutia	2825.66	82.10	13.58	6.47	67.67	25.86
2	Mahammadpur	1268.97	63.04	25.63	9.38	63.38	27.25
3	Kendamari Jalpai	2174.62	61.39	18.35	2.62	74.91	22.47
4	Nandigram	1069.62	80.40	20.35	6.98	82.56	10.47
5	Haripur	1260.36	87.28	15.00	6.36	75.45	18.18
6	Gokulnagar	2102.38	83.71	14.49	3.98	84.66	11.36
7	Samsabad	1208.88	80.24	12.89	5.15	84.54	10.31
8	Daudpur	1142.11	93.69	19.16	43.46	47.20	9.35
9	Kalicharanpur	3112.31	40.32	12.35	2.39	77.69	19.92
10	Sonachura	2022.3	65.77	12.78	2.26	75.19	22.56
	TOTAL	**18187.21**	**70.38**	**15.74**	**8.09**	**73.49**	**18.42**

Source: Agriculture Development Office, Nandigram

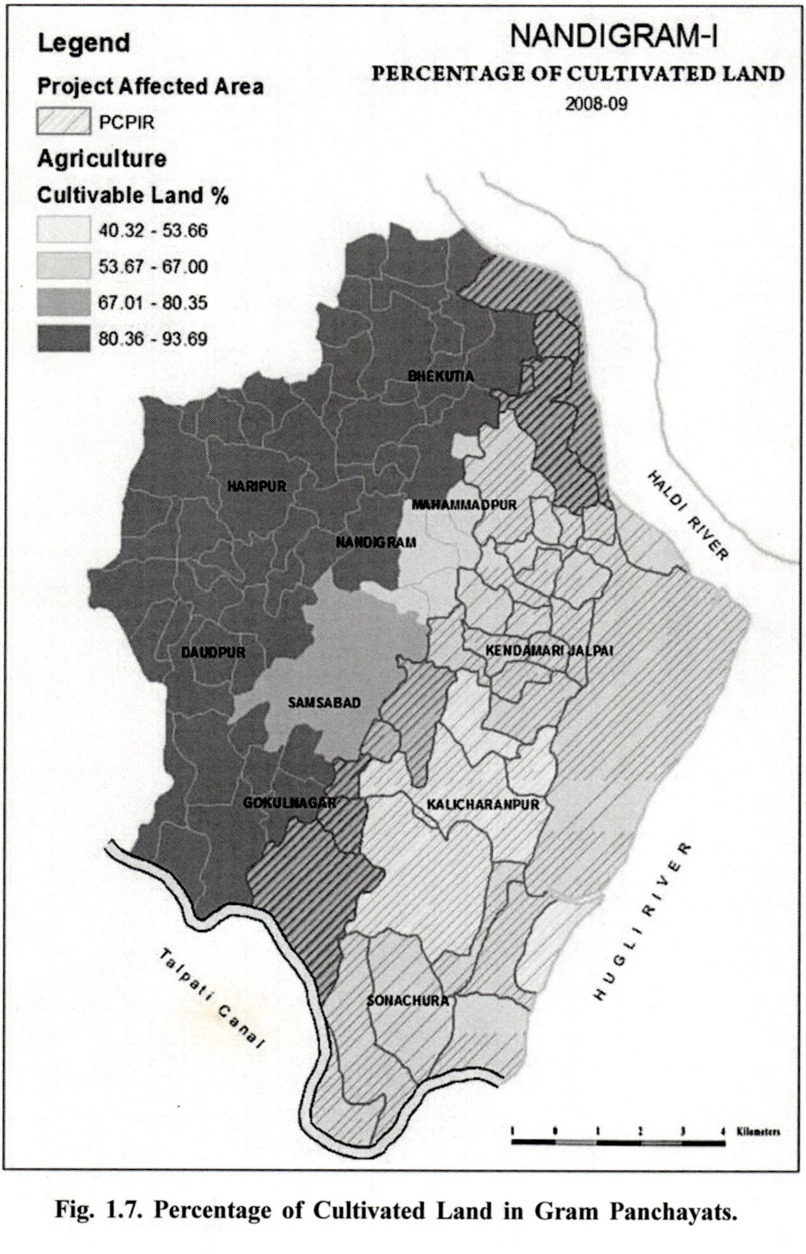

Fig. 1.7. Percentage of Cultivated Land in Gram Panchayats.

Table 1.4: Crop Coverage Report of All Crops in Nandigram-I (GP wise in hact.) 2008-09

GP_ID	*Gram Panchyat*	*Aus Paddy*	*Aman (Local)*	*Aman (HYV)*	*Kharif Veg.*	*Elephant foot*	*Ginger*	*Termeric*	*Colocasia*	*Kalai*	*Khesari*	*Moong*
1	Bhekutia	7	337	1821	82	10.25	9.25	9.75	6.25	15	630	23.5
2	Mahammadpur	1	100	654	27	5	5	5	5	10	370	15
3	Kendamari Jalpai	1	125	1126	40	6	5	5	6	15	570	18
4	Nandigram	0.5	80	715	40	2.75	4.5	4.5	3.75	10	170	11
5	Haripur	6	235	770	45	9.5	8	8.5	10	15	370	21.5
6	Gokulnagar	10	250	1390	77	5	6	5	7	10	270	13
7	Samsabad	4	100	800	46	3	3	3	5	6	250	7
8	Daudpur	4	195	78	40	7.5	6	6	8	12.5	270	14.5
9	Kalicharanpur	4	175	1000	48	9	5	5	5.5	9	485	11
10	Sonachura	6	150	1100	43	6	6	6	5.5	10	570	9
	TOTAL	43.5	1747	9454	488	64	57.75	57.75	62	112.5	3955	143.5
	2007- 08	12	1390	10495	410	31	22	36	0	50	3000	125
	2008- 09	16	1800	10182	500	50	54	30.25	0	84	3975	50

GP-ID	*Gram Panchayats*	*Sunflower*	*Mustard*	*Til*	*Wheat*	*Winter Veg.*	*Potato*	*Boro Rice*	*Summer Veg.*	*Betgel vine*	*Others*
1	Bhekutia	12.5	10	0	1.75	75	70	75	64	13.65	0
2	Mahammadpur	10	10	0	1	30	40	120	28	8	1
3	Kendamari Jalpai	10	10	0	0.5	50	60	130	27	15	1
4	Nandigram	7.5	8	0	1.5	40	50	50	27	5.5	0
5	Haripur	9	9	0	2	45	55	120	23	16	17.5
6	Gokulnagar	15	5	0	3	75	75	90	45	15	14
7	Samsabad	9	3	0	1	45	40	50	25	7	7
8	Daudpur	12	7.5	0	2	40	40	180	20	15	20
9	Kalicharanpur	16	5.5	0	2	50	35	30	12	13	0
10	Sonachura	12	5	0	1.5	50	35	35	22.5	20.5	13.5
	TOTAL	113	73	0	16.25	500	500	880	293.5	128.65	74
	2006- 07	81	80	18	12	540	490	895	260	118	12
	2007- 08	15	46	35	12	500	500	900	260	118	12

Source: Agriculture Development Office, Nandigram-I.

Crops and Cropping Pattern

The annual crop coverage statistics of 2008-09 generated by the Agriculture Development Office, Nandigram Block-I reveals more than 20 varieties of different crops grown regularly in this area (Table 1.4). Those include cereals, pulses, oil seeds, tubers, vegetables and betel vine, etc. The data is also plotted on a map to show the percentage and spatial variation patterns of the crops in all ten GPs (Fig. 1.8). The size of the circles in the map is proportionate to the cumulative area under all crops. Hence, considering the multiple cropping the cropping intensity is computed as 130 per cent in the whole Block.

High-yielding variety *Aman* paddy dominates heavily among all crops in all parts of the Block followed by *Khesari* (a type of pulse) which is grown in the same *Aman* paddy fields and harvested afterwards. Even local verities of *Aman* paddy are grown in wide areas. On the other hand, production of potato and vegetables, both in summer and winter, are showing the growing recent trends, which include several HYVs fetching good cash to the farmer. It is remarkable to observe that the traditional cash crop jute has completely disappeared from this area mainly due to low market price.

It is further interesting to observe that there is hardly any perceptible variation in crop coverage between the river side and inland areas indicating almost equal value of agricultural returns in whole of the Block.

The current agricultural crop calendar, as above, presents four complete seasons along with the main season for the fast growing pisciculture. Other than the paddy dominated seasons, the recent introduction of several vegetables and potatoes have extended the seasons. The pre-*kharif* season is a new trend here with the availability of quick growing species. In fact, it is evident that there is hardly any agricultural fallow for most of the plots in the Block.

Historical Backdrop

The detailed history of this region is beyond the scope of this present report. However, a brief account of the historical and political backgrounds of this region is given below for understanding the

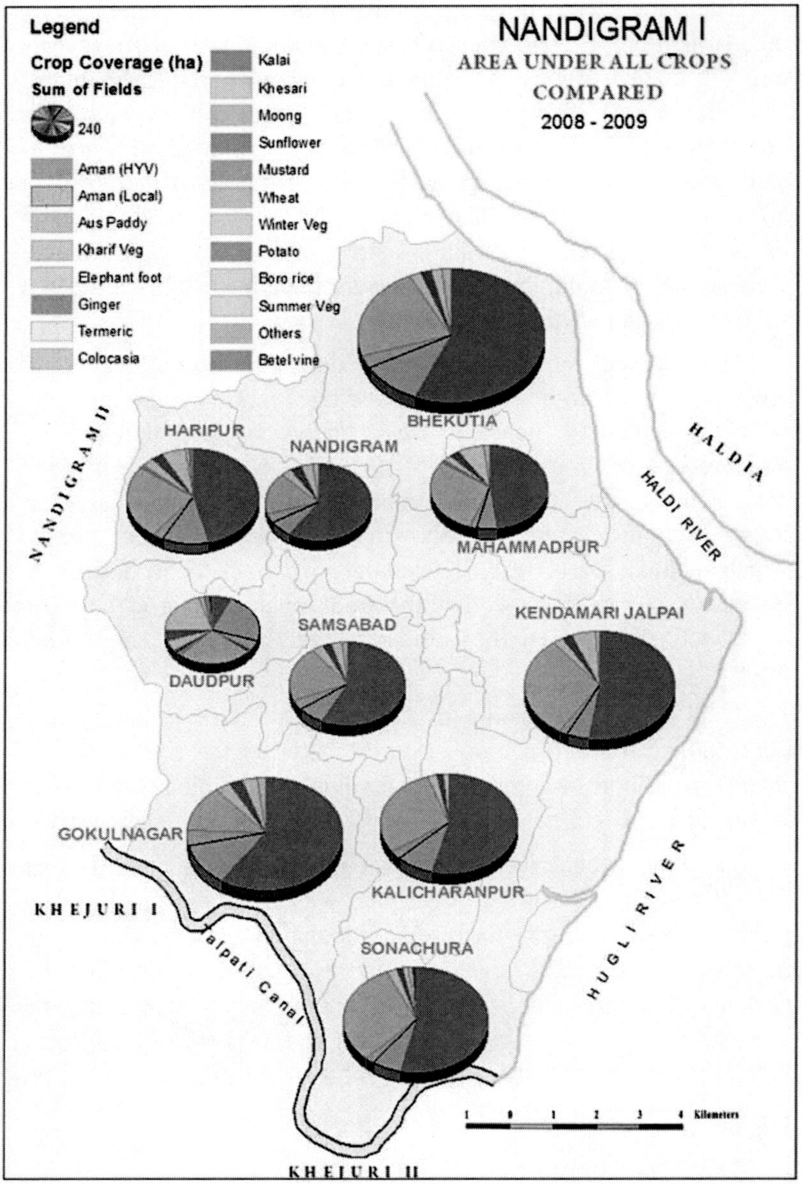

Fig. 1.8. Cropping Pattern in Gram Panchyats.

Table 1.5: Agricultural Calender of Nandigram Area

<table>
<tr><th colspan="12">AGRICULTURAL CALENDER OF NANDIGRAM AREA</th></tr>
<tr><td>JANUARY</td><td>FEBRUARY</td><td>MARCH</td><td>APRIL</td><td>MAY</td><td>JUNE</td><td>JULY</td><td>AUGUST</td><td>SEPTEMBER</td><td>OCTOBER</td><td>NOVEMBER</td><td>DECEMBER</td></tr>
<tr><td></td><td></td><td></td><td></td><td></td><td></td><td></td><td></td><td></td><td></td><td></td><td></td></tr>
<tr><td></td><td colspan="4">Summer Crops
[Boro paddy, bitter gourd]</td><td></td><td colspan="5">Kharif
[Aman paddy, til, brinjal, Vindi, Khesari etc.]</td><td></td></tr>
<tr><td colspan="3">Winter (Robi)
[Til, Brinjal, Cauliflower, Chili, Potato, vindi, Sweet potato]</td><td colspan="4">Pre Kharif
[Mung, Kalai,Chili, Sugarcane, Bitr.Gourd(pebble]</td><td colspan="4"></td><td>Winter
(Robi)</td></tr>
<tr><td colspan="6"></td><td colspan="4">Pisciculture</td><td colspan="2"></td></tr>
</table>

Source: Field Data.

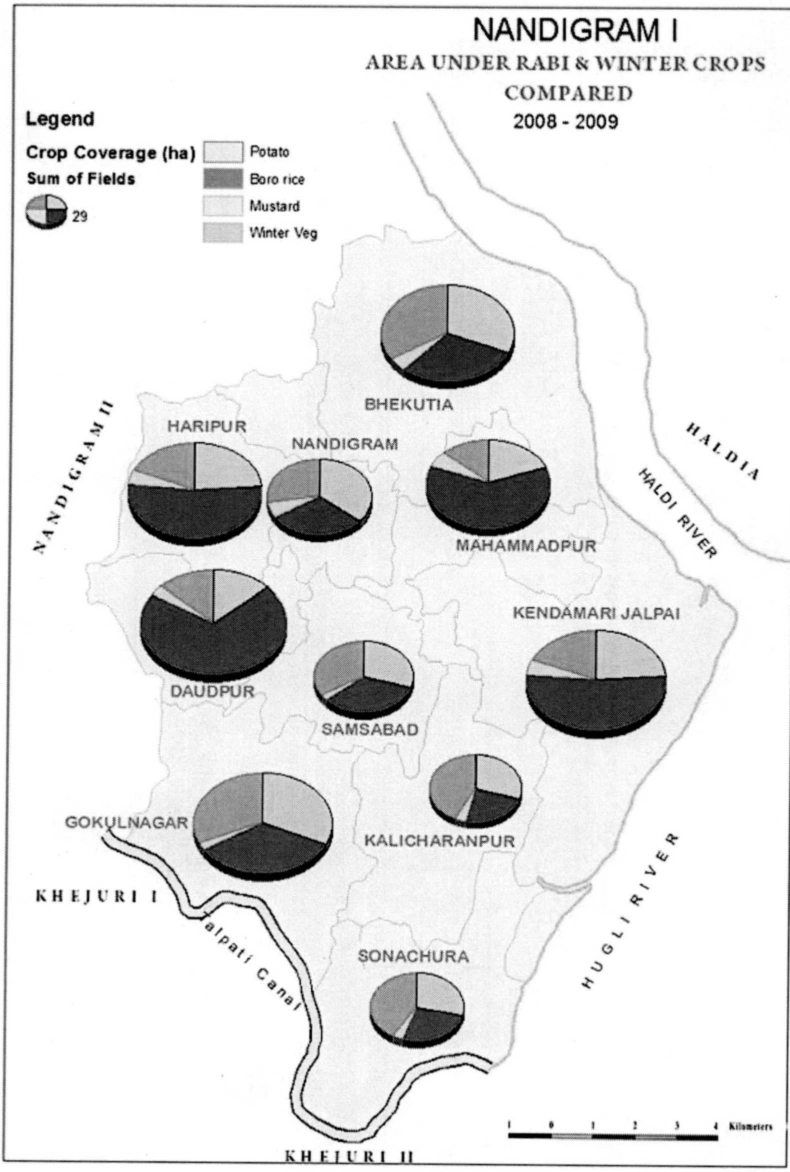

Fig. 1.9. Distribution of Ravi and Winter Crops in gram Panchayats

present socio-economic and political scenario. This region has a glorious past as evidenced from the inscription of *Tamralipta*. In ancient times, various invasions were done especially the invasion by Emperor Asoka (Circa 261 BC). The famous Chinese traveller Fa Hien visited *Tamralipta* (640 AD) and wrote an interesting account about the social and economic conditions of the people of this area, especially of Tamluk.

During this long period many events took place and had reacted on the minds of the people living here. These had influenced their activities, life cycle, social and religious rites and obligations and contributed largely towards the development of a new culture.

Medinipur district witnessed a series of revolutions. Especially 1799 AD is marked as the year of the great *Chuar rebellion* or the revolt of the *Paiks* (country-soldiers). This rebellion took a serious turn and some of the British *sepoys* and the British collectors were killed by the *Paiks* (Price, 1874).

Nandigram is the history of struggle for freedom since the days of the British rule in India during the latter part of the ninteenth century. During the 1942 Movements, this district was freed from the British administration and villagers established a parallel government at some police stations and hold on continuously as independent units for few months. The people, irrespective of their castes and creeds, joined this glorious revolution and boldly faced bullets and imprisonment. The people of Nandigram also witnessed a few more of movements and these have reflected the unstable condition of the agrarian people who belong to many castes and communities within a wide range of social hierarchy. The agricultural share-croppers launched a movement popularly known as *Tebhaga* movement in1946-47 in this region under the leadership of legendary communist Bhupal Panda. The share-croppers (*bargadars*, *bhag chashis*, *adhyars*) of Bengal began to assert that they would no longer offer half the share of their crop to the *Jotedars* or landlords but only one-third. The area had also experienced the *barga* movements by the peasants to fight against the local landlords. The assimilation of the above mentioned two kinds historical facts led the local people into a strong mental makeup which is anthropologically termed as 'Frontier'[2] (Bhowmick, 1976).

In 1982, another movement under the leadership of Bhupal Panda originated in Nandigram with the demand for development (roads, sanitation, water supply, electrification, etc.). Police fired on the agitation and killed a student, Sudipta Tewari. At that time, village-folks dug trenches in Nandigram to prevent police from entering into the villages.

Besides, such spirit of movements, other natural calamities like cyclones, famines, etc., which this area faced from time to time also brought the people closer to each other on humanitarian grounds and led them to unite on a common goal of social and political oneness. In fact, this part of Bengal never witnessed any communal violence with a share of 60/40 population between the Muslims and the Hindus.

Jellingham Project

As a part of industrial development, the Jellingham project of Burn Standard Industries Limited at Gangrachar *mouza* in Nandigram Block-I has been wearing out since its close down in the year of 1989. Close to 400 acres of land at Gangrachar was acquired in 1977 for this project by the Kolkata Port Trust and Burn Standard. About 142 families lost their land. Of them, 19 families had lost the roof over their heads. Some were lucky; they received a small compensation for their lost land after knocking on the doors of the Calcutta High Court. The production started in 1984 and after functioning for five years only, this unit of Burn Standard wound up all its operation due to mismanagement and conflict with labour unions. A huge area of this abundant project is still lying deserted and unattended. Although 142 farming families lost their livelihood during the land acquisition process at that time, only five got jobs in the factory but not for long. The outcome of Jellingham project has raised apprehension and mistrust among the resident of Nandigram in which also had bearings on the proposed plan for industrialisation.

1.5 Genesis of the Present Movement in Nandigram

The controversy started when the State government planned to build a chemical hub at Nandigram as part of the proposed Petroleum, Chemicals and Petrochemicals Investment Region (PCPIR) around 2006. Before December 2006, a rumour was there in Nandigram for nearly a year that some of the *mouzas* or villages might be acquired by the State Government for instituting an industrial area under Special Economic Zone (SEZ).

The genesis of this problem further cropped up regarding a special circular issued by the Haldia Development Authority on 28 December, 2006 [Haldia Development Authority (HDA) as B-XVII-HDA Notification dated 28.12.2006]. The circular stated that land in following 27 *mouzas* of Nandigram Block-I and II *mouzas* of Khejuri Block would be acquired for the proposed chemical hub:

	Name of the Mouzas	*J.L. No.*
A.	**Nandigram Block**	
1.	Dinabandhupur	165
2.	Madhabpur	186
3.	Parbatipur	187
4.	Nayachak	188
5.	Kendamari-jalpai	189
6.	Bainchibari	190
7.	Hajarakata	191
8.	Nilpur	192
9.	Hosainpur	193
10.	Bhasurkata	194
11.	Suvanichak	196
12.	Mirzachak	197
13.	Rajaramchak	198
14.	Shyamsundarchak	199
15.	Osmanchak	200
16.	Jadubarichak	201
17.	Garchakraberya	202
18.	Jalpai (7 NO.)	246
19.	Kalicharanpur	248
20.	Saudkhali-Jalpai	249
21.	Gangra	250
22.	Sonachura	251
23.	Sonachura-Jalpai	252
24.	Gangrachar	253
25.	Soudkhalichar	254
26.	Nakchirachar	255
27.	Charkendamari	250
B.	**Khejuri Block**	
1.	Sahebnagar	47 and 48

Source: Haldia Development Authority Notification 'Vide No. B-XVII; dt. 28.12.2006.

Chronology of Events

A public meeting was held on behalf of the then ruling party CPI(M) in Nandigram market on 29th December, 2006 where the then MP from Tamluk, Purba Medinipur district urged farmers to pave the way of development through industrialisation in Nandigram out of surrendering their lands, both farmland and residential land, against compensation. Hearing the news, the male-folk returned to their villages with certain apprehensions. At the same time, a large number of male-folks, predominantly Muslims, came to their villages for *Idujjoha* on 1st January, 2007, who were working in Metiabruz, Kolkata or elsewhere. They also became agitated hearing the news that their birth-place (*matribhumi*) is going to be forcibly occupied by the government.

(a) Incidents of January 2007

The trouble in Nandigram began with attacks on Panchayat members, administrative officials and police on 3rd January, 2007. On that day, a preparatory meeting for the visit of a central team for declaring the area as 'Nirmal' Panchayat, for excellent achievements in sanitation was taking place in Kalicharanpur Gram Panchayat office. Some activists of the then opposition parties gathered there at around 11.30 in the morning. They demanded that the land acquisition notice, served by Haldia Development Authority be scrapped and the panchayat declare, that there would be no such acquisition. Samerun Bibi, the then panchayat *pradhan*, declined to pay attention to their demand. Demonstrators became violent and abusive. They ransacked the panchayat office. The panchayat secretary was injured in this attack. The mob also pelted stones on the Garchakraberya health sub-centre. It was only after this that the panchayat *pradhan* informed the Nandigram police station. As the police vehicle was proceeding towards the village, they were attacked by an armed mob. It is reported that eleven policemen including two assistant sub-inspectors were seriously injured. The police vehicle was torched. After some time, another mob attacked a police car of Khejuri police station about five kilometres away and attempted to burn it. The police did not enter the area since then [*A Statement of Facts* by CPI(M) Party, March 16th, 2007].

The mob started moving with arms and began to destroy bridges and culverts linking roads. On 4th and 5th January, they virtually destroyed every link road and bridge connecting Nandigram and Khejuri to the outside world. Villagers reorganised the barricades and trenches on roads in a manner so that one could move by walking, bicycle or even on *rickshaw*, but no car or speedy motorbikes could move in or out. Big trees were also placed across approaching roads so that police vehicles could not enter inside the villages. The people started night guard for constant vigil towards the situation. This type of 'calm and cool' condition in the area was a short time breathing space.

The ferry service connecting Nandigram and Haldia through Haldi River was cut off temporarily, the supply of cash crop and labour from Nandigram to Haldia was halted. Haldia market faced a huge price rise at that time. Nandigram and Khejuri are separated by a 50 ft wide canal called Talpati Khal. This bridge separates two newly constructed *pucca* roads (under *Pradhan Mantri Gram Sadak Yojana* or Prime Minister Village Road Project, 2004-2005) namely, Talpati Khal to Basulichawk (16.2 km) in Nandigram and Talpati Khal to Rasulpur in Khejuri. In this context it is to be mentioned that historically, Nandigram was attacked time and again by the Portuguese *Armada* (*Harmad*[3] in colloquial dialect) or pirates and British invaders through Khejuri. Nandigram people used to resist them. During the present movement, a large number of CPI(M) workers were forced to leave the villages and took shelter in a nearby camp.

On 4th January 2007, to protest against the notice issued by the Haldia Development Authority thousands of people gathered in front of Garchakraberya under the banner of *Gana Unnayan-O-Jana, Adhikar Sangram Samity* and *Nandigram Jami Uchchhed Birodhi- O-Jana Shakti Raksha* Committee. Nandigram continued to remain cut off from the rest of the state, several roads had been dug up and some bridges and culverts also were damaged. Thereafter, the Trinamool Congress, Socialist Unity Centre of India (SUCI), *Jamiat Ulema-e-Hind* and Indian National Congress cooperated to establish the *Bhumi Uchchhed Pratirodh* Committee (BUPC, Committee against eviction from land). The first BUPC formed on

5th January 2007, in a meeting held at *Etimkhana* (Islamic boy's home) located at the Johurimore of Nandigram with the following members representing different political parties:

Sl. No.	*Name of the Member*	*Resident of*	*Political Affiliation*
1.	Subhendu Adhikari	Contai	TMC leader
2.	Abu Taher	Kanchan Nagar, Kendemari	TMC (formerly Congress)
3.	Sheik Sufian	Nandigram	TMC (formerly Jela Parishad Purta Karmadhyakhsa)
4.	Nanda Patra	Nandigram	SUCI leader
5.	Sheik Samad	Daudpur	PDCI leader
6.	Sheik Kushnabir	Samsabad	CPI(M) (Panchayat Pradhan)
7.	Sayam Kazi	Kalicharanpur	TMC
8.	Sheik Sahauddin	Kendemari	Formerly CPI(M) leader
9.	Swadesh Das	Gokulnagar, Tekhali	Panchayat Member-Gokulnagar G.P.
10.	Advocate Piyus Bhuinya	Dhanyakhola	SUCI leader

Source: Field Data.

A large number of supporters of the then ruling CPI(M) party also joined them. The apparent aim of the BUPC was to protect the farmers' lands. However, the top leadership of the then ruling party determined to ride over all opposition against the industrialisation project. Because the then ruling party argued that the land reforms have considerably ameliorated the economic condition of the rural mass, and was prepared to take on industrialisation for which strong domestic market has already been created (Guha, 2007).

Nandigram was again attacked on the onset of 7th January, 2007, from Khejuri, with bombs and bullets. It was a foggy morning. Thousands of people from the adjacent villages (Bhangabera, Sonachura, Gangra, Adhikaripara, Gokulnagar, Tekhali, etc.) started thronging at Talpati Khal Bridge, to resist the invaders. Three persons died of bullets during resistance, namely, Bharat Mondal, Sheik Selim and Biswajit Maiti (12 years old).

It is reported that on 7th January 2007, that at least five persons from Nandigram were killed and some of the injured persons were transferred to SSKM Hospital in Kolkata.

(b) Incidents of 14th March, 2007

On the 14th morning, on the occasion of *Chaitra Sankranti* thousands of villagers assembled at three places–Bhangabera (near Talpati canal on the Nandigram-Khejuri border), Gokulnagar (near Tekhali bridge) and Saraswati Bazar (in Garchakraberya) to offer holy prayers.

This was, of course, part of an ongoing series of popular nonviolent protests and resistance strategies for the preceding two/three months to oppose the proclaimed government decision on acquisition of productive farm land from Nandigram Block-I. The movement was organised by the villagers themselves spearheaded by the local leadership under the banner of the BUPC.

The villages by the side of the Talpati canal were mostly populated by Hindu community. And the heart of the resistance was Garchakraberya, a Muslim dominated village, which is 10 km away from the Talpati Khal. The *shankhadhwani* (blowing sound of conch shell) by Hindu women-folk started as early as at three am on 14th March. Children and women gathered beside the Talpati bridge at three a.m. and started *puja* of *Gouranga* idol and *Singhabahini,* the community believes that the goddess save their land and lives. Muslim women and children came a little late, at four a.m. and started reading from Koran Sharif. Thousands of villagers, mostly children and women were present there since the dawn. No prominent leaders of BUPC were available at the spot, but local village level leaders were there.

The police force along with some other members of political supporters, as stated by many villagers, entered through Talpati canal into the territory of Sonachura-Jalpai village and firing was initiated by the police force to disperse the gatherings of villagers of surrounding villages. Tear gas cells were also thrown towards them and many villagers were affected by this incident and caused eye troubles in the form of eye irritation. During field investigation and while taking of household census, it was observed that most of them who had assembled in that gathering experienced eye trouble and still they are suffering from it despite treatment provided by local doctors and doctors of Nandigram hospital.

In that fateful morning of 14th March, 2007, in Nandigram Block-I of Purba Medinipur district of West Bengal, 14 villagers (according to official account) died and about 60 people injured in a firing by the police on thousands of villagers including hundreds of women and children. The violence continued throughout the day and the following day. According to the claim of the then state administration, the objective of police intervention was mainly for the restoration of damaged roads. The firing and bombing claimed 13 lives in Nandigram on that day whose names are listed below:

Sl. No.	*Name*	*Age*	*Father's/Husband's Name*	*Village*
1.	Imdadul Khan	20	Abdul Daihan Khan	Jalpai(NO. 7)
2.	Gobinda Das	30	Bhanucharan Das	Jalpai(NO. 7)
3.	Badal Kumar Mondal	45	Gobardhan Mondal	Jalpai(NO. 7)
4.	Supriya Jana	40	Sukumar Jana	Sonachura
5.	Jaidev Das	24	Haradhan das	Sonachura
6.	Rakhal Giri	32	Pratap Giri	Sonachura
7.	Panchanan Das	45	Gunadhar Das	Keshabpur
8.	Sambhu Pal	30	Rabindranath Pal	Keshabpur
9.	Ratan Das	30	Kanai das	Gangra
10.	Pushpendu Mondal	27	Pulin Mondal	Gangra
11.	Raja Seikh Imdadul	20	Seikh Imdadul	Jadubarichak
12.	Basanti Kar	62	Gorachand Kar	Kalicharanpur
13.	Pralay Giri	25	Sarbeswar Giri	Saudkhali

Source: Field Data.

The dead individuals are from the villages like seven No. Jalpai, Sonachura, Keshabpur, Gangra, Jadubarichak, Kalicharanpur and Saudkhali. Besides, many individuals had been injured on that day due to police firing and tear gas. The villagers sought treatment from the local practitioner, Nandigram hospital, Tamluk hospital and some of the serious patients had been referred to the S.S.K.M. hospital of Kolkata. The villagers suffered a lot physically, mentally, socially and still they are suffering from the trauma which they had expressed during field investigation in that area.

Meanwhile, a district-wise 12 hours strike was called at the afternoon of 14th March by CPI(M) Trade Union wing CITU, to prevent the protesters from hampering the process of reinstallation of administration in Nandigram, according to *'The Hindu'*, a media source. Sonachura, Bhangabera, Adhikaripara, Tekhali were completely under the control of police and 'armed CPI(M) cadres' by the night of 14th of March (*Manthan Samayiki*, Thursday, April 5, 2007).

(c) After 14th March, 2007

Under the direction of Kolkata High Court, CBI team arrived at Nandigram on 15th March, 2007. Even after 14th March, 2007 the presence of police camps in the area, located mostly in the local schools, has a negative impact on the education and welfare of children. The entry points to Nandigram area were all sealed by the protesters denying access to the others and even to the media. Several writers, artists, poets and academicians took a strong position against the police firing which in turn brought significant international attention.

In April 2007, a People's Tribunal on Nandigram was set up by a group of concerned citizens from different parts of India who felt the need for an independent investigation into the violence of 14th March, 2007. The initiative was inspired by a call given by Justice Krishna Iyer, Mahashweta Devi, M.T. Vasudevan Nair, Rajendra Yadav and Ashok Vajpeyi for the restoration of peace and justice as well as to ensure to all those whose rights have been violated in any form. Following members constituted the People's Tribunal on Nandigram: (1) Justice S.N. Bhargava, Retd. Chief Justice, Sikkim High Court, (2) Prabhash Joshi, Founder Editor, *Jansatta,* (3) Lalita Ramdas, Social Activist, (4) John Dayal, Journalist and Human Rights Activist, (5) Dr Jyotirmay Samajder, Psychiatrist. The venue and date of the said Tribunal was at Gokulnagar Gobindajiu

Prathamic Bidyalay at Sonachura, Post Nandigram, Purba Medinipur, West Bengal on 26th and 27th May, 2007 and at University Institute Library Hall, Kolkata on 28th May, 2007. On each day, the Tribunal resumed at 10:30 a.m. and continued until 5 p.m.

(d) Incidence of November 2007

On 6th November, 2007, the police post near the Tekhali Bridge was withdrawn. It is also reported in the newspaper that the peace procession at Gokulnagar of Nandigram was attacked by the followers of the then ruling party with arms ammunitions and six people were absconding since then (*Anandabazar Patrika*, 25th May, 2012, p. 7).

On 10th November, 2007 a peace rally was called by BUPC. It was alleged by BUPC supporters that firing was initiated by CPI(M) activists to destabilise the procession. The BUPC accused the then ruling party for large scale violence by which they abducted many persons to use these people as shield to enter inside the villages to retain the control there. After 10th November, 2007 the entire Nandigram was in turmoil state as a war like situation between BUPC and CPI(M) activists. This episode was continued for consecutive days. Afterward Central Para Military Force was deployed in Nandigram to control law and order. In the meantime, peace was restored and many people were able to come back to their villages with help of Central Para Military Force.

On 8th May, 2008 the Panchayat election took place with active participation of 'Central Reserve Police Force (CRPF)' for a free and fair poll. Most of the informants are of opinion that CRPF had a great role to maintain peace in Nandigram. They had been deployed in strategic points where different groups of people could assemble, for instance market or source of drinking water, etc. The 2008 Panchayat election was the turning point in view of shift of power equation. CPI(M) lost seven Gram Panchayats and TMC came in power by wining Nandigram Panchayat Samity and Purba Midnapur Zilla Parishad. TMC further strengthened their position by winning the Parliamentary poll in 2009 and West Bengal Assembly election in 2011.

Table 1.6: Nandigram Assembly Election results since 1977

Year	*Winning Candidate*	*Vote Perce-ntage*	*Party*	*Losing Candidate*	*Vote Perce-ntage*	*Party*
2011	Phiroja Bibi	60.17	All India Trinamool Congress	Paramananda Bharati	34.75	Communist Party of India
2006	Illias Mahammad Sk.	48.71	CPI	Sk. Supian	45.33	AITC
2001	Illiyas Mahammad Sk.	51.5	CPI	Sunil Baran Maiti	45.4	AITC
1996	Debi Sankar Panda	49.23	INC	Sakti Bal	49.12	CPI
1991	Sakti Bal	47.85	CPI	Debi Sankar Panda	44.33	INC
1987	Saktiprasad Bal	52.62	CPI	Debi Sankar Panda	46.15	INC
1982	Bhupal Panda	54.34	CPI	Ramesh Chandra Gharai	42.85	INC
1977	Prabir Jana	44.62	JNP	Bhupal Chandra Panda	20.52	CPI

Source: Election Commission of India.

Now, the BUPC is almost inactive and except TMC all other partners like SUCI, PDCI, etc., has left the organisation. The Nandigram episode attracted the attention of judiciary through many court cases. The civil society was very active at that time. They regularly visited the Nandigram and were in constant touch with the local people. The print and electronic media played a proactive role at that time by supporting or opposing the BUPC. At present, there is an apparent peace prevailing there. The fear and disbelief are still perpetuated by the people. Many persons are still absconded; some CPI(M) activists are away from their own villages. A large number of victims got compensation according to court order. Even many others are waiting for said compensation. Many persons are still in fear of arrest because of the pending police cases against them.

Table 1.7: Khejuri Assembly Election Results since 1977

Year	*Winner name*	*Vote Percetage*	*Party*	*Losing candidate*	*Vote Percetage*	*Party*
2011	Ranajit Mondal	53.11	AITC	Asim Kr. Mondal	43.34	SP
2006	Swadesh Patra	57.82	WBSP	Dr Partha Pratim Das	39.38	AITC
2001	Sunirmal Paik	54.68	WBSP	Ramchandra Mondal	40.16	AITC
1996	Mondal Ramchandra	55.51	CPI(M)	Paik Sunirmal	42.08	INC
1991	Sunirmal Paik	53.21	CPI(M)	Santiram Das	41.48	INC
1987	Sunirmal Paik	58.16	IND	Mondal Susanta	38.9	INC
1982	Sunirmal Paik	53.44	IND	Mondal Sushanta	45.21	INC
1977	Sunirmal Paik	43.76	JNP	Sunil Sit	22.96	CPI(M)

Source: Election Commission of India.

In view of the attempt of the then West Bengal Government for introducing Special Economic Zone (SEZ) policy in Nandigram area, in the Panchayat elections of May 2008, CPI(M) and its left front allies were defeated in Nandigram and adjoining areas by the Trinamool Congress-SUCI alliance. After people's struggle the proposed Nandigram SEZ was scrapped and the PCPIR project has been declared withdrawn.

1.6 Objective and Relevance of the Study

In view of the above backdrop, a multi-disciplinary team comprising physical anthropologist, cultural anthropologist, ecologist and psychologist from the Anthropological Survey of India, Kolkata has conducted a study on the most affected villages of two Gram Panchyats, namely, Sonachura and Kalicharnpur of Nandigram-I during the period from 21st November–14th December, 2011. The broad objectives of the present study are:

1. To understand the general socio-economic and health profiles of the area;
2. To ascertain the local resource base and its utilisation;
3. To know the ethnic composition of the area and their intercommunity relationship;
4. To assess the people's perception about various development programmes;
5. To understand the need and priorities of the area; and
6. To ascertain the psycho-social stress and disorder, if any.

1.7 Selection of Villages for the Present Study

Although the entire Nandigram area was disturbed due to proposed land acquisition, but five to six villages were mostly affected, namely, Sonachura, Garchakraberya, Jalpai No. 7, Kendamari, Gokulnagar and Amdabad. For studying the recent socio-political scenario of Nandigram in connection with this movement two villages have been selected for the present study. One of those is Sonachura (J.L. No. 251) under the Sonachura GP, the southernmost villages of Nandigram Block-I, which was considered to be the epicentre of Nandigram unrest. This village is located about 20 kms south of Nandigram Police Station. This is a traditional Hindu village comprising 936 households with a total population of 5,305 individuals (2,709 males and 2,596 females). The other is the Muslim concentrated village namely, Garchakraberya (J.L. No. 202) under Kalicharanpur GP the first place where the violence erupted. This village comprises 599 household with a total population of 3,715 individuals (1,839 males and 1,876 females-Census 2001).

The Sonachura Gram Panchayat is surrounded by canals like Talpati canal (on south), Chandakhal (on north-east), river Hugli (on east) and Kalicharanpur Gram Panchayat No. 9 (on north). A metalled road has gone through this grampanchayat from north to south which has crossed the Talpati canal and gone to the Khejuri Block on southern tip of this grampanchayat. The incident of 14th March, 2007 took place almost at the junction of Talpati canal and Sonachura Jalpai where villagers decided to prevent the attack from the side of Khejuri through Talpati canal and started the religious performances of different religious groups.

1.8 Methodology and Limitation

The methodologies were worked out in consultation with inter-disciplinary team and utilised their respective tools and techniques like household census, case studies and interview, height and weight measurements for nutritional assessment of the children and psychological attributes, etc., in order to generate the empirical data in this field survey. Utmost care was taken with regard to objectivity and authenticity of the study, right from the survey design to filling up the questionnaires, data processing etc. The door-to-door field survey was conducted with 1,362 individuals in two villages, e.g., Sonachura and Garchakraberya—the most affected area. The identification of the victims was done on the basis of discussions with a cross-section of villagers, relevant hospital, PHC and medical camp reports. When we entered the villages as researcher, initially the villagers did not speak out their life-experiences but gradually many of them took us into their confidence level and expressed the causes of their involvement into the processions and agitations, directly or indirectly. It was a difficult task because of magnitude of the impact and at the same time the sensitivity of the issue. But after focused group discussion at various levels, the situation was very much congenial for collecting data and they voluntarily participated in this survey.

To address the socio-political issues regarding the incidents of Nandigram, a door-to-door household survey was conducted mainly in three villages, i.e. Sonachura, Sonachura-Jalpai and Garchakraberya. The households were randomly selected for this investigation and altogether 275 households were studied considering the age, sex, marital status, educational status, occupational status etc. of each of the members of the studied household along with the householdwise land-holding pattern. In this household survey, the direct and indirect involvement of the household members in the incident and the injuries and tortures were recorded during household enumeration.

Nature of Data

In this study, we have collected both the secondary data mostly the census data like area, population, sex ratio, etc., as well as the primary data generated from two selected villages. Both qualitative and quantitative data were collected. The agricultural data like cropping patterns, crop coverage and crop yield was compiled from Block Agricultural Development Office. The BLR Office of Nandigram provided the data on landholding and related land use map, etc., those macro level data were analysed and mapped on a GIS platform using satellite imageries and base maps to understand the spatial distribution patterns.

2

Land and Land Use Pattern in the Context of Present Movement

2.1 Changes in Land Tenure System and Land Reform

This part of the estuarine land in question, which forms almost 20 per cent of the total Nandigram Block-I, became functionally habitable and cultivable at a much later stage through the tidal process of siltation and stabilisation. These relatively recent alluvial deposits remained saline and water logged crisscrossed with smaller distributaries and creeks offering subsistence to mainly the fishermen, boatmen and cow herders.

During the 'Permanent Settlement' (*Sthayi Bandobasto)* under the British rule, agricultural farmers received huge areas of land (*Chakra*) in this newly developed riverine islands (*Nadichars*). Mahishyas dominated among those new owners (*mahajan).* The Maity, Jena, Sen families from Takapura village of Nandigram Block-II and different places in Khejuri police station area, Das family of Kalicharanpur and Sri Indra Maity of Mahishadal were the then big Jamindars and *mahajans*. Those *mahajans* were all non-resident owners and a 50:50 *barga* system was in practice for a longer period. While the West Bengal Bargadari Act, 1950 provided for a sharing of output in the ratio 60:40 if the sharecropper provided the inputs, it did not provide any security of tenure of the *bargadar*. Later on, the more comprehensive West Bengal Land Reforms Act, 1955 came into force, which mainly provided for the abolition of intermediaries and the imposition of ceiling on landholdings. There were amendments to this Act, whereby the right of cultivation of a *bargadar* was made heritable (1970), there were more stringent requirements for termination of *barga* contracts, and the *bargadar's* share was made 75 per cent (1:3) with input contribution (1965). However, since there was no complete official recording of *bargadars*, many of these rights were not realised in practice.

These reforms took shape mainly in the form of: (a) redistribution of vested land, and (b) securing of tenancy rights, which already existed in law, through a programme of universal registration of tenants called 'Operation Barga'. The West Bengal Landholding Revenue Act (1979) and the Revenue Rules (1980), introduced by the Left Front government, provided for key changes in the sharecropping system. These were in addition to two other means of land reforms that were undertaken in most other parts of India including West Bengal, namely, imposition of ceiling on large landholdings and the reduction of sub-infeudation through the abolition of intermediaries between the cultivator and the landlord.

The major change in tenant relations involved the active recording or registration of sharecroppers who had cultivated on the same piece of land for a number of years. This registration with the Department of Land Revenue gave them permanent and inheritable rights to cultivate the land. Though this legal provision already existed, it had not been implemented properly. In the late 1970s, the Government launched a serious drive to effectively implement the rights of the sharecropper. In particular, now the onus of disproving a claim to *barga*dari rights was very clearly put on the landowners'. The objective was to provide security of tenure to the sharecropper so that first, his/her livelihood is not threatened and second, he/she is encouraged to make permanent and more expensive improvements on land and adopt more modern technology. In addition, some recognised land title was recognised to be a precondition for access to formal agricultural credit. All this was obviously intended to increase the productive capacity of the land, which could then yield a stream of higher incomes for both the tenant as well as the landlord. In addition, the earlier legislation that provided for a percentage share of output in the ratio of 25:75 between the landlord and tenant (when the tenant bore the entire cost of production), and a percentage share of 50:50 when the landlord bore the costs (except for labour), was now sought to be implemented properly (SHDR, West Bengal, 2004).

The registration under 'Operation Barga' also entitled the tenant to subsidise institutional credit and other support facilities in the form of cheaper inputs like seeds, fertilisers and water. Indeed, there was a positive attempt to support the programme by increasing on the ground access to credit and fertiliser in the early eighties.

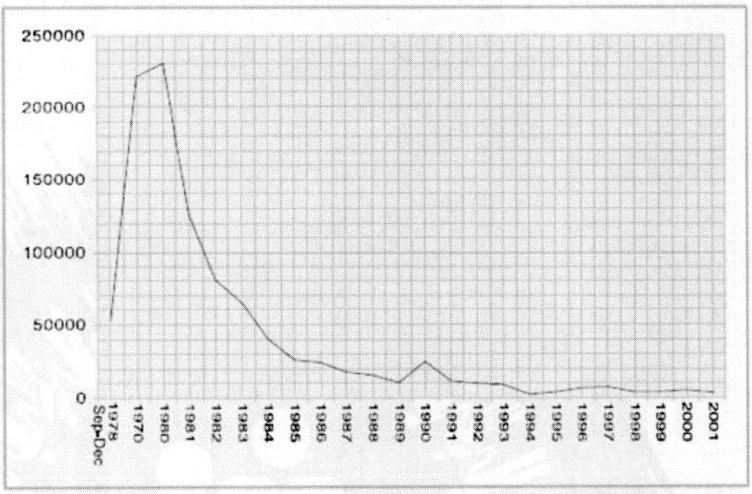

Source: Directorate of Land Records and Surveys, West Bengal.

Fig. 2.1 Progress in Recording *Bargadars* in West Bengal

In addition, as Chakraborti *et al.* (2003) have pointed out, over time this issue has also become much less of a concern among district officials, panchayats and peasant organisations which have been more involved with other issues. With an active thrust on the recording of tenants no longer a major policy focus, the registration appears to have become fairly routine, and, therefore, very sparse in terms of numbers. This may also reflect changes in land relations and contracts such that the issue of sharecroppers' rights is no longer the dominant concern. For example, the growth of fixed rent contracts for rice cultivation may also have contributed to reduce interest in *barga* registration. It could be that the very empowerment of tenants that was the initial aim and result of *barga* registration has led to a situation over time where the registration itself is seen to be less necessary or important (Table 2.1).

Table 2.1 District-wise Recording of *Bargadars* in West Bengal

District	*Number of Bargadars Recorded in 2000*	*Per cent of Total Cultivators in 2001*
Darjeeling	12879	15.6
Jalpaiguri	61279	23.4
Koch Behar	84431	23.4
Dinajpur	103717	22.4
Malda	81171	29.3
Murshidabad	84426	23.1
Birbhum	111654	43.1
Bardhaman	131691	38.2
Nadia	64022	20.5
North 24-Parganas	73946	27.3
Hughly	112996	25.7
Bankura	111654	25.4
Purulia	9071	2.6
Medinipur	292112	27.8
Haora	42717	6.1
South 24-Parganas	112939	31.7
Total West Bengal	**1082193**	**30.3**

Source: SHDR West Bengal 2004.

There have been recent reports of increasing land alienation by *pattadars*, and of eviction of *bargadars*, thus suggesting that the benefits of the land reform have been relatively short-lived at least for some rural households. Alienation is reported to be more marked in Darjeeling because of the urban extension of Siliguri, in Uttar Dinajpur as tea gardens seek to cover more area, and in parts of the Sunderbans area due to the expansion of brackish water fisheries. Eviction of tenants is also reported from areas of North Bengal in particular because these tendencies are potentially so significant; they are worth considering in more detail.

A recent study sponsored by the State Institute of Panchayats and Rural development (Chakraborti *et al.*) found that, on average, 13 per cent of the *pattadars* who received land had lost it by 2001. This is a relatively small percentage, given the original very small size of holding and the many forces that have made small peasant cultivation much less viable across India over the past decade. Of course, this percentage varies widely across districts, as Table 2.4 shows. The extent of land alienation of *pattadars* varies from a low of 5.62 per cent of *pattadars* in Medinipur, to a high of 22.35 (or more than one-fifth) in Uttar Dinajpur. Similarly, Table 2.1 shows that the northern districts are especially badly affected in terms of the proportion of *bargadars* who have been evicted.

It is also true that the very small holdings that were created by the distribution of '*pattas*', meant that there could be problems of economic viability in terms of the land providing full subsistence for a rural household. So other, or subsidiary, sources of income would have to be sought by the members of *pattadar* households. It could be that die evident expansion of non-agricultural economic activities in the rural areas of the state, along with the growing financial difficulties of cultivation, has created a situation where a focus on non-agricultural activities has become preferable for many small-holder rural households (SHDR, West Bengal, 2004).

Towards a better understanding of the land fragmentation due to land reform and redistribution in the form of *barga* and *patta* in the studied villages all plots (*dag* numbers) of one sheet each of the *Mouza* maps of Sonachuara and Garchakraberya have been taken as sample amounting to 378 such plots in total. The corresponding land records data have been extracted for those sample plots and classified according to land use and ownership (Tables 2.2, 2.3 and 2.4).

Table 2.2. *Barga* Scenario in Studied Villages

Village (No. of plots)	*No. of plots with Bargadars*	*Total land under Barga (acres)*	*Average size of plots (acres)*	*No. of plots with given patta*	*Total area (acres)*
Sonachura (N=178)	115 (64.61)	9.86	0.0857	1	0.03
Garchakraberya (N=200)	69 (34.5)	8.46	0.1226	8	1.67
Total (N=378)	**184 (48.67)**	**18.32**	**0.1000**	**9**	**1.70**

Source: DL & LR Dept. Purba Medinipur.

It is found that in total 48.67 per cent of plots are given to *bargadars* where it is almost double in Sonachuta compared to Garchakraberya. It is important that the average size of land under *barga* is only .0857 acre in Sonachura which are also relatively less fertile than that of the Garchakraberya having an average size of 0.1226 acre. Moreover, there is only one plot of land with *pattadar* compared to eight such plots in the respective villages. Therefore, it is found that the river side Sonachura is found less affected by *barga* than inland Garchakraberya in terms of total land under *barga* in contract to the number of plots and plot sizes. It should be mentioned here that there are many more unofficial sharecroppers or *bargadars* in the villages.

Table 2.3 Land Use of 200 Sample Plots in Garchakraberya Village

Category	*No. of Dag/Plot*	*Total Area (Acres)*	*Average Size (Acres)*	*Percentage to Total*
Moist Low Land	143	71.24	0.5	85.02
Homestead	9	2.24	0.25	2.67
Pond	10	1.14	0.11	1.36
Black Loam	13	2.04	0.16	2.43
Small Pool	5	0.35	0.07	0.42
Katcha Road	10	3.84	0.38	4.58
Khamar	1	0.01	0.01	0.01
Old Fallow	1	0.02	0.02	0.02
Dump yard	1	0.02	0.02	0.02
Cremation ground	3	0.11	0.04	0.13
Canal	3	2.74	0.91	3.27
School	1	0.04	0.04	0.05
Total	200	83.79	0.42	100

Source: Computed from DL & LRD data, Purba Medinipur.

In terms of land use categories, land under moist low land (*jaljami*) or the actual cultivable land amounts to above 80 per cent in both the villages. But black loam plots are almost double in the inland of Garchakraberya over the riverside of Sonachura. The striking difference is found in case of land under governemnt and village fisheries, among these two villages in Sonachura clearly leading the scene as expected. Average size of low land is also smaller in Sonachura (Tables 2.3 and 2.4).

Table 2.4: Land use of 178 Sample Plots in Sonachura Village

Category	*No. of Dag*	*Total Area (Acres)*	*Average Size (Acres)*	*Perc-entage to Total*
Moist Low Land	98	77.13	0.79	82.4
Govt. Fishery	1	4.62	4.62	4.94
Village Fishery	17	3.74	0.22	4
Homestead	17	2.35	0.14	2.51
Pond	8	1.65	0.21	1.76
Black Loam	9	1.4	0.16	1.5
Pucca House	2	0.87	0.44	0.93
Small Pool	12	0.65	0.05	0.69
Nala	3	0.31	0.1	0.33
Embankment	4	0.23	0.06	0.25
Old Fallow	2	0.18	0.09	0.19
Dhosa	2	0.17	0.09	0.18
Thana	1	0.16	0.16	0.17
Bamboo Grove	1	0.09	0.09	0.1
Cow Shade	1	0.06	0.06	0.06
Total	178	93.61	7.26	100

Source: Computed from DL & LRD data, Purba Medinipur.

2.2 Recent Trends in Agro-economy

In the deltaic south Bengal in general and Nandigram in particular, the *Barga* movement could not be sustained for a longer period mainly due to two reasons. The size of landholding with the *bargadars* and *pattadars* being very small and the poor farmers were not really economically empowered to adopt the newer technology and high yielding seeds towards a viable subsistence. Even the quality of land also hindered the farmers to develop a viable agriculture in certain areas. Such situations has gradually reverted the system towards growing dependency on moneyed farmers or *mahajans* and such small holders found no alternate than to illegally lease those lands for non-agricultural usage like highly profitable pisciculture and brick kiln, etc.

During the present study, it was reported by the villagers that there are a large portion of vested land still lying undistributed throughout the *Purba* Medinipur district where landless families are unofficially allowed to cultivate but *pattas* are not issued in their names leaving them as a kind of bonded to the local governance. Such group of population, living under a total uncertain economic status, is the most vulnerable group with a huge fear of getting no or little compensation from any land acquisition. They are also not eligible for any government development project meant for legal landholders. On the other hand, some *pattadar*s having no man or money power to cultivate the land and hence, they are selling the land against one time lump-sum.

The recent trend among the *bargadars* is to offer a deal to the owner for a 50:50 ownership of the land under *barga*. In such deals, the registration of the sale deed is fixed on the same date when *bargadar* first purchase the whole plot and then sell half of it to the legal heir of the owner bypassing any legal complications. The purchasers pay their respective registration fees, etc. and no money is actually being transacted. *Mahajan* find it profitable, mostly the non-residents, who cannot verify the actual production from the land to ascertain the one-third share.

It was also informed that almost all the *mahajans* of the riverside *mouzas* are from Nandigram Block-II area and have almost lost their direct control over the legal or unofficial *bargadars*. For them sale of the land under *barga* is a blessing in disguise and that such land under *barga* cannot be sold to any other person.

Another kind of temporary transfer of land is also in vogue in this area where owner in need of money mortgage a plot of land against a lump-sum for a year or two. The land is then cultivated by the lease holder for his own and that the leaser needs not to pay any cash interest for the loan amount. This is very similar to that of the long-term lease of land taken by the fishery owners.

Land Use Change

The trend of change in the land use offers a critical relation with the change dynamics of the land tenancy and ownership change. The current land use pattern and quantum of return from different land use categories ultimately determine the current value of land. This aspect is of immense importance to formulate mitigation measures through a well-accepted R and R Policy towards a successful land acquisition process.

From the Land use/Land cover map of 2005 developed by the Government of West Bengal was collected and discussed at the Block Land Reform Office, Nandigram-I and following interpretation has been done:

(i) Proposed SEZ area has been cropped out from the map and estimation is done in GIS platform to understand the percentage of land under different categories depicted in the map.

(ii) It is found that, contrary to the previous records that most of the agricultural land here is single cropped, there was approximately 38 per cent of the geographical area under double crop while 48 per cent produce single crop and about 18 per cent was covered with rural settlement (Fig. 2.2).

(iii) In fact several ravi and winter crops including boro paddy are now being cultivated all over the Block. The dry farming of post kharif potato is emerging as one important cash crop other than the mustard and different vegetables (Fig. 1.9).

(iv) It is to be noted that, as informed by the BLRO, the rural settlement area shown in the map actually includes small tank, pond, horticulture, and kitchen garden areas in and around the actual homestead land. Such tanks and ponds are also utilised for sweet water fisheries for both subsistence and commercial purposes. The domestic land, which is termed as High Land and locally, known as *Kalajami* is the best quality soil where profitable betel leaf vines are cultured.

(v) Within the single crop land, on the other hand, falls the lowest category low land or *Jal-jami*. Such land plots located in between the embankment and the river are at present being intensively utilised for highest return from prawn fishery (Fig. 2.3).

(vi) Therefore, though the 2005 land use map is giving an updated land use of the area but the current ground realities need to

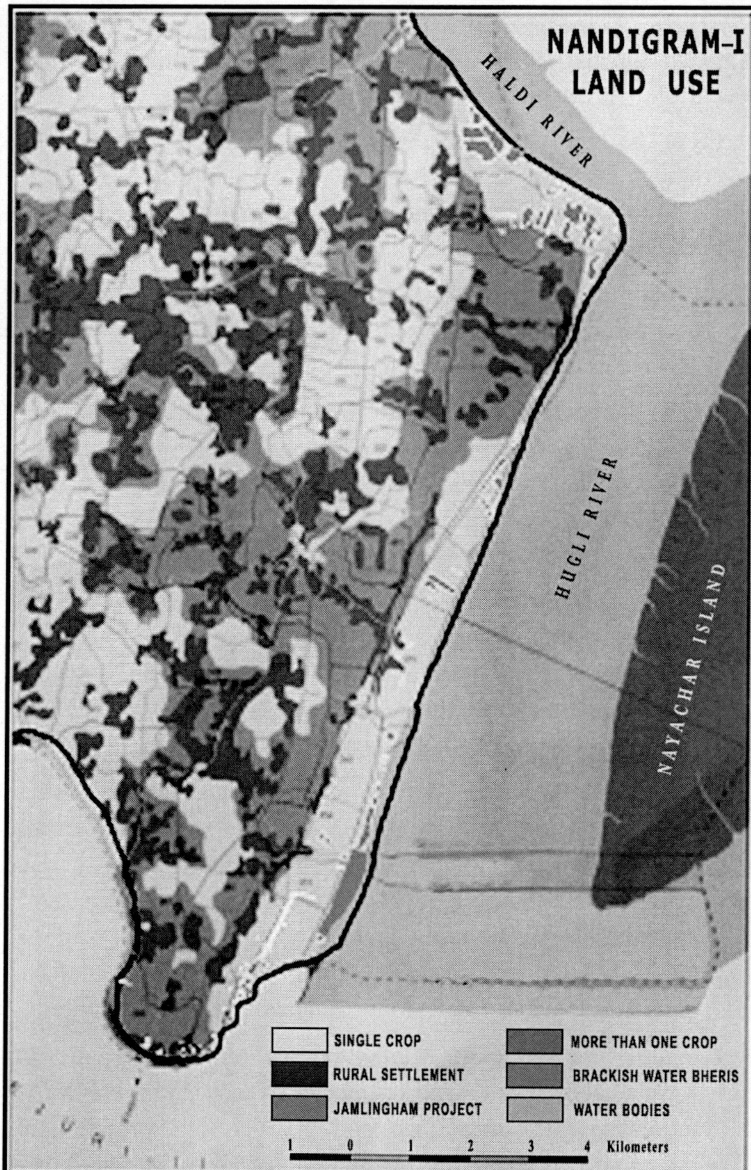

Fig. 2.2. Newly Emerged Brackish Water Bheris (Google Map) plotted on land use/land Cover Map 2005.

be added and the map should be updated to include such details of recent development, *e.g.,* the fishery ponds and other smaller water bodies, etc.

Fig. 2.3 Examples of Cultivable Land Increasingly Converted to Brackish Water Bheri as seen in Google Imageries taken during 2006, 2010 and 2014

Tiger-prawn Culture

In fact, pisciculture and other fishing activities in this estuarine ecosystem are as old as the civilisation itself in the sweet water dugout ponds, *bils* and rivers. The introduction of tiger-prawn culture and marketing of prawn fingerlings from the tidal rivers are relatively recent phenomena. Sri Ashok Mondal of Sonachura- Jalpai village in the Nandigram Block-I was reported by many villagers as the pioneer in introducing the culture in the whole of Purba Medinipur belt back in 1997-98. He came in contact with one Asraf Ali of Golabari in Barasat area of North 24-Parganas where this culture was very popular a few decades before. That time they used to sell fingerlings over a vast area covering far and wide locations like Kanthi, Haldia, Mayna, and even Odisha and Bangladesh.

Mondal with his younger brother remembers their first successful harvest in 1999 which they did in the low land owned by panchayat *pradhan* taken on lease. Primarily, technicians from IFB used to come for any support to the beginners and in case of any disease or infections. Now a days several such companies send their technicians from Kanthi, Digha or even from Visakhapatnam.

From detailed interview and on site observations the following deductions can be made on the highly popular and profitable fishery activities and technicalities:

(a) Traditionally fishing is practised in the creeks, canals and *aman paddy* fields primarily for subsistence.

(b) Sweet water fisheries (white water fisheries) also have a long tradition in the region in the community or personal ponds and *bils* in a commercial scale. Fishes like *Rui, Katla, Mrigel, Pangas, Gule, Bele, Tangra, Pankal, Parse* and small verities of prawn are cultured here with traditional system and have a very good local and even inter-district market. Fingerlings are available mostly from big hatcheries 'Ramganga' in Bankura and also from Kanthi area or even from localities. A net profit of 100 per cent plus is reported per harvest on an average per year, which is done in the fifth month.

(c) The tiger-prawn or *Bagda* culture is the latest trend mostly done along the Hugli and Haldi riverside lands and on either side of the river canals and creeks, which supplies ready brackish water along with certain natural nutrients from the tidal rivers. The fishes like *Bagda* and *Bhetki* are main varieties in these *bheries*' the fingerlings of which are naturally available from the tidal rivers. The most affected *mouzas* are in the five riverside Gram Panchayats, i.e., Bhekutiya, Mahammadpur, Kendamari Jalpai, Kalicharanpur and Sonachura. In total, 148 such active fishery ponds are located in the Google map imagery accessed online which are located mostly in the east of the main embankment running parallel to the Hugli River.

(d) Such fishery ponds or *bheries* are prepared by mechanically digging the land in a proper rectangular form on any category of land say low land, wetland, old community fisheries or good cultivable lands which are adjacent to the source of tidal water as only rainwater is not considered suitable for this. Generally, two-three feet of top layer is dug and the soil is used for raising embankment of one-two feet so that at least four feet deep water can be reserved for the fishery.

(e) The villagers having provisions for investment of good amount of money generally go for such *bheri* business which is considered a high capital, high risk and high return venture. Particularly, the small and marginal landholders in the riverside as then less fertile areas are readily giving an instant cash return against their land. This year in Gnagra Char *mouza* per *bigha* lease price of such land rose to Rs 20,000 per *bigha* per year and the lease agreement was done for minimum five years subject to payment of 50 per cent at the time of agreement in cash and the rest in third or fourth year on in annual instalments.

Changes in Tiger-prawn Culture

The ponds are now dug deeper to hold five feet to 5.5 feet deep water instead of three-four feet in earlier times. Earlier, there was no second crop harvested in the autumn. The culture has now become mostly dependent on special prawn feed both from India and abroad and dependency on several medicines has grown from earlier years. Seedlings are now coming mostly from Andhra Pradesh and Uttar Pradesh well preserved in ball packs. The production has also been almost double in two decades with more investment and better marketing options.

Legal Aspects of Tiger-prawn Culture

While accompanying the BLRO of Nandigram Block-I during his visit to several *bheris* in Sonachura Jalpai village on 7th December, 2011 on a mission of mapping all the fisheries in that area on the existing land use map of 2005. Several interesting observations were made as follows (Fig. 2.2).

(i) The BLRO declared and appealed to the *bheri* owners that the present government is eager to record all the fisheries in operation as there are huge increase in number of *bheri*es during last three-four years and that most of these fisheries are developed on either the vested land or on *rayat* land through illegal lease and in both cases the change in character of land have not been registered through conversion. Thus, while the owners are earning a lot the Government department is losing good amount of land revenue. The District Magistrate has taken up the matter personally and if the owners fail to act within this December the authority will take strict necessary measures.

(ii) For land under Mechhogheri, the occupants must appeal for short term lease against payment of minimum *Selami* to the government renewable every five years. Those *bheri* owners had taken illegal lease of cultivable land from *rayaties* should produce application for conversion from each *rayati* or owner of the land. Those developed *bheri*es on vested lands of other category or having *barga* or *patta* should also apply either for short-term (five years) lease and/or for conversion.

(iii) On the other hand, the district level officers in the Directorate of Fisheries at Tamluk also narrated that primarily this department is not against the new trend of brackish water fisheries in the area as a huge number of landless and marginal farmers are being benefited and this option has radically changed the economic level of the people of these areas.

(iv) In fact, if farmers develop fisheries or take part in any kind of fishing activities, viz., brackish water, marine or river water and in sweet water, taking proper permission from the Fisheries department the fees for land conversion will be less than those doing privately.

From the above observations and information, it is found that the State Land and Land Revenue Department is as such not going against practice of pisciculture but wants to get those legalised for the first instance. But by rule no *bargadar* has legal right to convert the land or sub-lease, which stands as a huge legal problem for this kind of fisheries speedily developing. This is obvious that such a condition brews conflict between the farmers and the government authorities to the extent that even the local officers can't dare to make official visits to those areas like Nakchira Char in the Nandigram Block-I for surveying and recording those illegal *bheri*es.

Case Study of an Experienced Farmer

One Sri Rabindranath Maity, aged 79 years, Mahishya by caste, a fourth generation resident of Sonachura Jalpai village adjacent to the Talpati Khal was interviewed on 4th December, 2011 afternoon while seated under a tree on the embankment road and knitting a throw-net meant for fishing. He possesses six *bigha*s of medium and low agricultural land (*Jaljami*) along with some high Land (*Kalajami*) and a piece of homestead land of about eight *cotta*. He lives here with the family of his younger son. His elder son was separated and two daughters are married. He also has a cattle shed on the bank of the khal where there are five number of cattles and six goats other than poultry birds. One Pan Boroj (betel vine) of about 2.5 *cotta* and a good number of areca palm, coconut, jackfruit, mango, etc. in his orchards and a medium size pond.

His great grandfather, a resident of Takapura now in Nandigram Block-II, was primarily a boatman and fisherman but later started sharecropping in the *mahajan*'s land. He first bought land in this riverside area from one *mahajan* (Mahisya caste) in Takapura. His father, who actively participated in *swadeshi* movements, particularly in the 'Laban Satyagraha', and had several times been chased and beaten up by the British police as people of this area used to produce good salt as a profitable occupation. His father started paddy cultivation as one of the pioneer. Several varieties of paddy were grown even in slightly saline soil with a good monsoon rain, e.g., *Patnai* (blackish and course), *Kamal Bhog*, *Dudher Sar,* etc. but *Khesari* and *Kalai* cultivation was absent. That time it was 50:50 share of crop.

Sri Maity, himself a Mahishya, actively participated in the Tebhaga movement, at his 20-22 years of age, against the Mahishya dominated *mahajan* group. But due to devastating flood waters during his grandfather's time, most of the land purchased by his great grandfather became un-cultivable and was sold out. His father bought land again in the same village which is inherited by him. He believes in cultivation of paddy, potato, pulses and vegetables mostly for his own sustenance. He sells some paddy after a good harvest as well as coconuts, areca nuts etc. He has not yet opted for *bagda bheri* though he agrees there is a good return and foresees a huge increase in such fishery during 2012. He is literate and his son passed middle school and a farmer. His grandsons and granddaughters are studying in school but lost a year due to this violent movement.

At the end, he narrated sadly that he has seen and participated in bigger political and peoples movements like that for India's independence and farmers' or tenants' rights over land but the recent one was the bloodiest and unimaginable one. But he and his family members neither ever left their home nor shifted somewhere away from this embankment road towards lower land to cover themselves from bullets and shells fired from Khejuri on other side of the Talpati *Khal*.

3

Socio-cultural, Economic and Political Dimension

3.1 Socio-cultural Dimension

Socio-political processes take its' own courses through conflicts and confrontations. The reasons of conflicts may vary from one social situation to another social situation depending upon its nature and involvement. Sporadic conflicts are part and parcel in every society but a society as a whole had experienced a situation due to a controversy over the issue of land at Nandigram. "Politics may broadly be seen as the establishment of authoritative decision through an agency in a society or a larger nation" (Das, 2011). That authoritative decision may not be accepted by a larger group of the society which may lead to resentment. These resentments may be capitalised by opponent party members and their leaders. Any decision whether political, economic, social or otherwise works on the society and it may work for the betterment of the society for some group of people or it adversely affects a section of the society.

For real understanding on socio-cultural aspects attributed by political dimension there is a need to highlight main emphasise of the study area which are firstly, to understand the reasons of peoples' unrest and present socio-economic and political state of Nandigram. Secondly, the after effect of the large-scale violence was evaluated based on case studies by maintaining academic neutrality.

Socio-religious Profile

Historically, this place had socio-religious dichotomy. The first one is the strong existence of local cultural traits manifested by temples and certain customary practices and second one is to combat the internal and external forces from time immemorial. The assimilation of the above mentioned two historical facts developed the local people into a strong mental makeup which is anthropologically termed as 'Frontier'[3] by Professor P. K. Bhowmick. The detail discussion on cultural traits of this region is not possible here. Despite that, few examples are important. There are three temples in Sonachura village. *Basuli Mayer Mandir* in south, *Singhabahini Mandir* in east and *Holi Mandir* in Golpara. Each temple had distinctive characters. There are some pebbles and stones inside the temple in case of *Basuli Mayer Mandir*. They believe these stones or pebbles could move once upon a time. Baikuntha Mondal, Gunadhar Samanta and Khirod Maity established the *Singhabahini Mandir* in 1335 *Bangabdda* (almost 85 years ago). Mother goddess *Durga* is worshiped in this temple. The *Holi Mandir* is also an age-old temple. Now, all temples are segmented due to political intervention. The dominant communities are Mahishya and Paundra Kshatriya in Sonachura and Jalpai villages respectively. They believe mainly in *Baishnav Sect*. Due to political cleavage there are two *Durga Pujas* celebrated one in *Singhabahini Mandir* and another one in Sonachura market. The people of the village also worship the goddess *Manasa*.

The Muslims in this area are socially and politically close to local Hindu population also. The informants in Garchakraberya village said in many occasions that there is no past record of riot in Nandigram. There were many old bandits in their community. At the time of turmoil, these old bandits were converted into aggressors. The above mentioned social and cultural factors are responsible for involvement of the people in fierce fighting and build up the leadership.

The ethnic compositions of the studied villages are presented in Table 3.1. Altogether eight Hindu caste communities as well as Khan Muslim and Sheikh Muslim as religious groups are reported. Mahishya (31.22 per cent) and Paundra Kshatriya (24.10 per cent) are the predominant communities in the area and concentrated in Sonacura village while 33.22 per cent of the total households are

Muslim and concentrated in Garchakraberya village. Kaibarta, Rajbanshi and Dhopas are distributed in both the villages.

Table 3.1 Ethnic Compositions of the Studied Villages

Ethnic Group	*Sonachura*	*Garchakraberya*	*Total*
Muslim	0 (0.00)	102 (87.93)	102 (33.22)
Mahishya	96 (50.26)	0	96 (31.27)
Paundra Kshatriya	74 (38.74)	0	74 (24.10)
Namasudra	11 (5.76)	0	11 (3.58)
Kaibarta	5 (2.62)	5 (4.31)	10 (3.26)
Napit	1 (0.52)	0	1 (0.33)
Rajbanshi	1 (0.52)	6 (5.17)	7 (2.28)
Daule	2 (1.47)	0	2 (0.65)
Dhopa	1 ().52)	3 (2.59)	4 (1.30)
Total	**191 (100.00)**	**116 (100.00)**	**307 (100.00)**

Source: Field data.

Table 3.2 Location-wise Community Distribution and their Occupation

Location	*Community*	*No. of household*	*Cultivator*	*Agricultural labour*	*Daily wage labourer*	*Service*	*Fishing*	*Tailoring*	*Betel vein*
Sonachura	Mahishya	96	55	1	9	9	2	0	2
	Paundra Kshtriya	74	34	10	12	6	1	0	2
	Namasudra	11	9	0	0	2	0	0	0
	Kaibarta	5	0	4	0	0	0	0	0
	Napit	1	0	0	0	0	0	0	0
	Daule	2	1	0	1	0	0	0	0
	Rajbanshi	1	0	1	0	0	0	0	0
Garcha-kraberya	Muslim	102	19	13	2	4	1	12	0
	Kaibarta	5	0	0	0	0	0	0	0
	Rjbanshi		4	0	0	0	1	0	0
	Dhopa	4	2	0	0	0	0	0	0

Source: Field data.

Another dimension is the social network of the local people, which is an important parameter to determine the unity of the people in any kind of political movement. In this regard, we have collected the marriage affiliation of the community to obtain a pattern of social relationship and direction which might help to communicate and unite people over a large area for a common cause. Table 3.3 reveals that majority of the communities are maintaining a marital alliance within the Block particularly Nandigram-I (71.98 per cent) followed by the adjoining Block Khejuri (13.44 per cent). It also indicates that the communities, namely, the Mahishya, Paundra Kshatriya, Kahn Muslim and Sheikh Muslim have spread their marital alliance even outside the district.

Table 3.3 Community-wise Marital Affiliation

Community	*Nandigram- Block-1*	*Nandigram- Block-2*	*Khejuri*	*Remaining part of Midnapur*	*24- Parganas (South)*	*Other districts of W.B*	*Other states*
Mahishya (n=134)	94 (70.14)	3	25	5	4	1	2
Paundra Kshatriya (n=106)	74 (69.81)	4	18	2	5	0	3
Khan Muslim (n=49)	34 (69.38)	2	2	6	2	2	1
Sheik Muslim (n=44)	34 (77.27)	2	1	1	4	1	1
Namasudra	8 (80.00)	0	1	0	0	1	0
Kaibarta	6	0	0	0	0	0	0
Mali	3	0	1	0	0	0	0
Napit	2	0	0	0	0	0	0
Rajbanshi	2	0	0	0	0	0	0
Total (n=357)	**257 (71.98)**	**11 (3.08)**	**48 (13.44)**	**14 (3.92)**	**15 (4.20)**	**5 (1.40)**	**7 (1.96)**

Source: Field Data.

3.2 Economic Dimension

Local Resource Base Availability and its Utilisation of the Studied Villages

Any kind of political uprising is determined by economic scenario of the region. The main reason is economic. The characteristics of the economic dimension are multifaceted. The first one is types of economic pursuits. The entire Nandigram area may be divided into three kinds of economic activities, viz., agriculture, betel leaf cultivation, fisheries and brick manufacturing. The paddy is the main crop and mostly rain dependent. Paddy cultivation is followed by pulse and potato cultivation. Three kinds of betel leaves are cultivated, viz., *Bangla Pan* (third grade and require more lime) followed by *Kali Pan* (second grade and testy) and *Mitha Pan* (first grade export quality, local people do not consume). Betel leaf is very sensitive, it requires due attention all the time. The field should be in highland. One row will be 50 ft in width constituting maximum 45 small creeping plants. Usually one betel leaf field locally called *Pan Baroch* yields 1,000 to 2,000 creeping plants. One farmer can earn Rs 5,000 to 7,000 per month by selling betel leaves. They prepare *Goch* for selling to local markets. One *Goch* constitutes 50 betel leaves. The local markets are in Mecheda, Tamluk and Kokotia in East Midnapur district. They can sell one *Goch* for Rs 10 to 12. There are varieties of betel leaves, for instance thin, fatty, small, big and medium size. The big and medium sizes are best one. The betel leaves cannot be lifted from here and there. The rule is strictly followed. The leaves just above the root would be rejected. The leaves in middle portion are highly demanded. The leaves of upper portion are non-sailable but can be consumed locally. The betel leaf plantation is very sensitive. The plantation is totally protected from scorching sun by thatching. The watering is made to the plants at the evening time.

Fishery is one important economic pursuit in this region. In Nandigram, two areas for example Gangra and Nakchirachar are famous for fisheries. Khejuri is another Block adjacent to Nandigram where large numbers of fisheries are there. Usually, low quality of lands is selected for fisheries. The second precondition is that the salty water always enters inside the fisheries and there will be the scope of discharge of salty water. Hence, the creeks are to be connected with the fisheries. High tide water either at the time of full moon or new moon are required for the fisheries. The problem is that sometimes sufficient water cannot enter inside fisheries at the time of high tide. The pump is required to collect the salty water needed. Normally 20,000 tiger-prawns are cultivated in one fishery where the profit money is Rs 60,000. Generally, within one year the tiger-prawn once cultured are exhausted. A tiger-prawn achieves full growth within three months. Sometimes, the rolling of water through machine is practised for production of 40,000 tiger-prawns in a year. After production of tiger-prawn or *Bagda*, the *Galda* prawn is cultivated, which requires six months' time for full growth. The profit in case of *Galda* prawn is not as high as in case of tiger-prawns. Though the fisheries derive very profitable income but require a good amount of capital, even risk factors are also important. Under that circumstance, most of the people give the fisheries in lease for Rs 25,000 to 30,000 for one year to ten years contract (with registration). Small companies are getting the contracts of these fisheries. Few farmers convert five-acre land into a fishery and employed a professional person for three months basis with a contract of Rs 20,000 with a profit of Rs 40,000. The fisheries are related to the growth of brickfields. The dugout soil from fishing ponds is mostly utilised in the brick kilns established at nearby location.

Land is the most important economic dimension particularly where land acquisition is the root of the problem. The landholding pattern is important. Since, 1977, particularly at the time of left front government regime the land reforms started. *Barga* operation and distribution of *Khas* (vested) lands to the marginal farmers are important steps taken by left front regime by which an enormous public support was generated. During our fieldwork, we have found that possession of agricultural lands is not restricted to certain families. The lands are segmented due to land reforms. Another important observation is that in Sonachura, Jalpai and Garchakraberya villages the number of *bargadars* (share croppers) is less. Most of the landless and marginal farmers are engaged in *Thika Chas* (contractual farming). *Thika Chas* is an annual agreement between owner of land and the cultivator. Initially, cultivator has to give an amount to the owner, then the cultivator takes the possession of that land and cultivate. The total yield will be under full custody of the cultivator. The cultivator can either sell the entire or partial yield to the market or consume. This agreement is valid for one year only. In the next year, the landowner can renew the agreement or discontinue. After a long time, the cultivators were experienced with land reforms and became capable to express their displeasure over non-receipt of *patta* (lease) of the vest lands they had acquired long back. Table 3.4 will give an idea on landholding of the Nandigram area.

Table 3.4 Community-wise Landholding Household (in *cottah*)

Community	*Landless*	*Up to 20*	*21-50*	*51-100*	*101-150*	*150-220*
Mahishya	28	22	15	8	3	1
Paundra Kshatriya	28	18	4	6	0	0
Namasudra	1	7	0	3	0	0
Khan Muslim	7	5	4	1	1	1
Sheik Muslim	29	5	2	3	0	0
Napit	1	0	0	0	0	0
Rajbanshi	1	0	0	0	0	0
Mali	3	1	0	0	0	0
Kaibarta	5	0	0	0	0	0
Total	**103 (48.36%)**	**58 (27.23%)**	**25 (11.73%)**	**21 (9.86%)**	**04 (1.88%)**	**02 (0.94%)**

Source: Field Data.

It can be revealed from the Table 3.4 that majority of the communities living in this area are landless or marginal farmer. It shows that the Mahishya and Khan Muslims mainly possess the land of about 100 *cottah* followed by the Paundra Kshatriya. From the above table, it may be concluded that no stratification on the basis of landholding could be ascertained because of the fact that no particular community possesses a big portion of lands in this area. Rather all communities are in landless position or marginal farmers. The next important dimension is seasonal migration specifically in respect of occupation and in search livelihood.

Table 3.5. Community-wise Occupation of Migrated Person

Community	*Private Service*	*Embroidery Work*	*Masonry Work*	*Daily Wage Labour*	*Tailoring Work*	*Others*
Mahishya	13	0	2	8	0	2
Paundra Kshatriya	2	0	4	14	0	2
Namasudra	0	0	1	0	0	2
Khan Muslim	0	1	0	1	5	0
Sheik Muslim	0	0	0	0	8	3
Napit	0	0	0	0	0	0
Rajbanshi	0	0	0	0	0	0
Mali	0	0	0	0	0	0
Kaibarta	0	0	0	0	0	0

Source: Field Data.

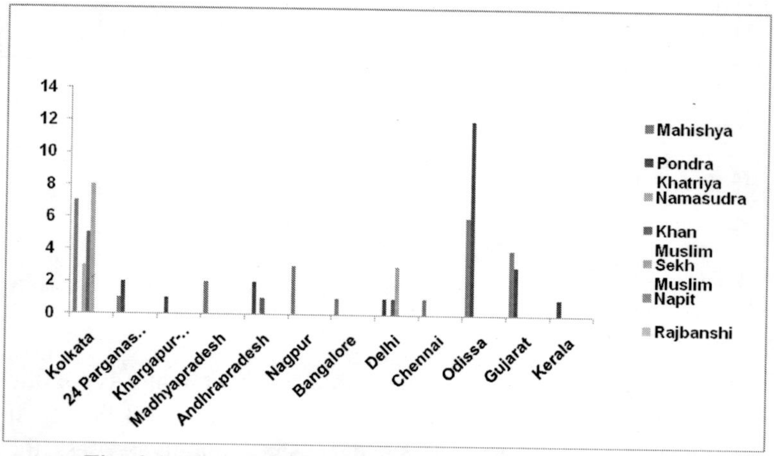

Fig. 3.1. Community-wise Migration from Nandigram

Source: Field Data.

This chart shows how different communities had migrated to different places. The Mahishya had migrated to Kolkata and Odisha. Paundra Kshatriya had migrated to Odisha, whereas Sheikh Muslim had migrated to Kolkata.

The migration is a common phenomenon in Nandigram. In most cases, one to three male family members used to go to Kolkata or different places of Odisha, Madhya Pradesh, Gujarat and southern states in search of job due to scarcity of job opportunity in and around Nandigram. They contribute a portion of their income to families who reside at ancestral villages. Usually, they return to their own villages at the onset of monsoon and harvesting seasons. Our field data on MGNREGA reveals that the jobs created by this scheme are not sufficient to meet their actual need.

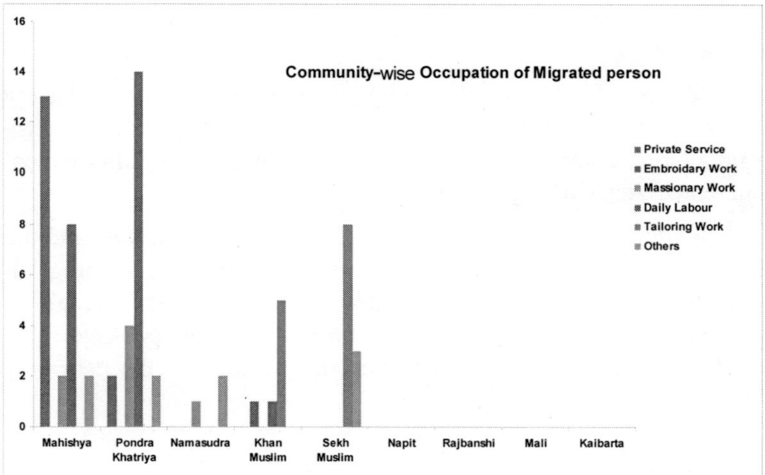

Fig. 3.2. Community-wise Occupation of Migrated Person

Source: Field Data.

This chart shows distinctive features on how different communities migrated for different occupations. The striking point is that the Mahishya are engaged in private service where the Paundra Kshatriya engaged in daily labour work and masonry work. Muslims are expert in tailoring work.

The possession of cattle is important parameter to determine the economic status. The following table will focus some idea on that aspect. Table 3.6 shows that all communities possess the cow largely. The number of bullock has decreased to a considerable extent due to introduction of power tiller and easy availability of that.

Table 3.6. Community-wise Cattle Census in Studied villages

Community	*Cow*	*Poultry Bird*	*Goat*	*Duck*	*Bullock*
Mahishya	74	7	25	0	2
Paundra Kshatriya	32	10	16	3	0
Namasudra	12	0	3	0	0
Khan Muslim	8	0	3	0	0
Sheik Muslim	12	0	7	0	0
Napit	0	2	0	2	0
Rajbanshi	0	0	0	0	0
Mali	0	0	0	0	0
Kaibarta	3	0	0	0	0

Source: Field Data.

We were able to collect data on community basis B.P.L. (Below Poverty Line) card holders (Fig. 3.3) and also community-wise beneficiaries under Mahatma Gandhi National Rural Employment Guarantee Act (MGNREGA) (Table 3.7).

The developmental programmes initiated by Panchayats and the State Government indicate certain trends. After the Nandigram episode in 2007 such developmental programmes was given momentum to compensate the after effects as far as possible. This trend could not able to continue and decreased over the period of time.

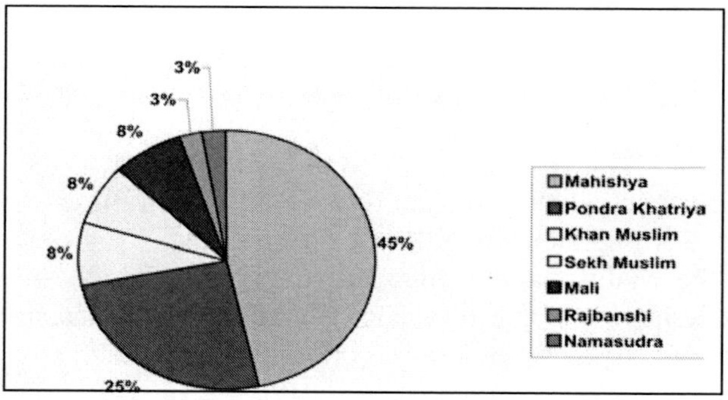

Fig. 3.3. Community-wise B.P.L. Card Holder

Source: Field Data.

This chart indicates that the families of Mahishya community possesses highest number of B.P.L. cards followed by Paundra Kshatriya, Khan Muslim and Sheikh Muslim. However, this figure is not absolute indicator of poverty in terms of ethnic dimension. In that context it is to be mentioned that Muslims are the most poverty stricken followed by Mahishya, Paundra Kshatriya and Namasudra in studied villages (Table 3.1)

Table 3.7 Community-wise MGNREGA Beneficiaries

Sl. No.	*Community*	*Year*	*Number of days employed*	*Daily Wage Rate*	*Total Amount*	*Nature of Work*
1.	Mahishya	2008-2009	55	75	4125	Digging of soil; plantation.
		2009-2010	184	81	14904	Cannel and pond cleaning; construction of dam; plantation.
		2010-2011	70	81/100	6772	Construction of dam; plantation and road repair by Morum.
		2011-2012	3	130	390	Construction of dam; plantation and road repairing by Morum
2.	Paundra Kshatriya	2008-2009	23	75	1725	Digging of soil; plantation.
		2009-2010	272	81	22032	Cannel and pond cleaning; construction of dam; plantation.
		2010-2011	64	81/100/130	5755	Construction of dam; plantation and road repairing by Morum.

contd...

contd...

Sl. No.	*Community*	*Year*	*Number of days employed*	*Daily Wage Rate*	*Total Amount*	*Nature of Work*
		2011-2012	13	130	1690	Construction of dam; plantation and road repairing by Morum.
3	Sheik Muslim	2008-2009	101	75	7575	Digging of soil; plantation.
		2009-2010	38	81/100	3306	Cannel and pond cleaning; construction of dam; plantation.
		2010-2011	12	100	1200	Construction of dam; plantation and road repairing by Morum.
		2011-2012	58	130	7540	Construction of dam; plantation and road repairing by Morum.

Source: Field Data.

The Table 3.7 shows that the progress of MGNREGA is satisfactory up to 2010-2011. By the help of that scheme construction of 'Morum' roads, cleaning of ponds, plantation was major works done.

3.3 Political Dimension

The political dimension is an important determinant of any kind of people movement across the world. Politicisation as referred by B.B. Goswami (1979) "to have discourse or on to engage in politics by the members of a society or groups of society to give a political tone or character or colour to the culture for supporting the decision taken" In the present context the entire unrest was rooted due to agony of land acquisition by state government for a proposed chemical hub under SEZ. To look into the problem it is important to go through the historical facts. During second left front regime, Nandigram witnessed a student movement with a demand to establish industry and development in Nandigram. This student movement was led by "Chhtra Parisad" (student union of Congress Party). In relation to that, it is to be mentioned the Trinamul Congress Party was not formed at that time. One student was killed due to police firing at the time of movement. On contrary, the opposition parties in 2007 demanded just opposite view to stop land acquisition even it was in favour of industrialisation. This incidence was categorically mentioned by a social scientist Professor Krishnendu Bhattacharya who resides more than three decades in that area at the time of interview with him. However, firstly there is a need to analyse the socio-economic background of leadership. The tables and charts will give an impression on that.

The people's version was that BUPC activists were actively supported by 'Maoists' while CPI(M) assembled their cadres from whole West Bengal. In this report, direct involvement means those who are actively involved in various political and other groups like BUPC (Bhumi Uchchhed Pratirodh Committee), Left Front, TMC (Trinamul Congress) and the like. In general discussions and discussions during household surveys, it has come up many times that Maoist influence also played a role during that time and they entered the area as members of a devotional singer group (*Kirtania*) and used to perform *Kirtan* (a kind of devotional song) moving from one household to another in that locality. Besides, they used to take food from some of the households from probable sympathisers there. Sometimes villagers were also compelled to provide them required inputs not fully aware of their background and motivations. This fact is also corroborated in some newspaper reports. The indirect involvement of these sympathisers also played a vital role in the movement.

Leadership

In any mass movement or unrest like this, the development of leadership is important. Sociologically the leadership is not imposed rather it is the inherent development from the society from which it is originated. The following discourse presents the socio-economic background of the leaders compared among the two contending groups in this area (Tables 3.8 and 3.9). The leadership of BUPC and CPI (M) are discussed at length. The study shows that leaders

Table 3.8 Social Affiliation of Leadership of BUPC

Sl. No.	*Name*	*Designation*	*Age*	*Sex*	*Education*	*Village*	*Community*	*Occupation*	*Political Affiliation*
1	Sheik Sufian	President	48	M	VI	Nandigram	Sheik Muslim	Business	TMC
2	Abu Taher	Joint Secretary	45	M	Graduate	Gopimohonpur	Sheik Muslim	Business	TMC
3	Nanda Patra	Joint Secretary	52	M	Graduate	Nandigram	Mahishya	Cultivation	SUCI
4	Pijush Bhuiya	Advisor	58	M	B.A., L.L.B	Dhanyakhola	Mahishya	Lawyer	TMC
5	Sayam Kaji	Local leader	44	M	VI	Kalicharanpur	Sheik Muslim	Tailor	TMC
6	Pranab Mahapatra	Local leader	45	M	Graduate	Sonachura	Paundra Kshatriya	Cultivation	TMC
7	Sheik Samad	Local leader	46	M	Maulana	Sarengabari	Sheik Muslim	Madrasa Teacher	TMC
8	Monoranjan Manna	Local leader	52	M	Secondary	Kalicharanpur	Mahishya	Cultivation	TMC
9	Sadesh Das	Local leader	45	M	Higher Secondary	Gokulnagar	Paundra Kshatriya	Cultivation	TMC
10	Meghnath Pal	Local leader	46	M	Graduate	Haripur	Mahishya	Cultivation	TMC
11	Late Nishikanta Mondal	Local leader	54 (age at death)	M	Secondary	Sonachura	Paundra Kshatriya	Cultivation	TMC
12	Banashri Khara	Local leader	64	F	Post Graduate	Nandigram	Namasudra	Retd. Teacher	TMC

contd...

contd...

Sl. No.	*Name*	*Designation*	*Age*	*Sex*	*Education*	*Village*	*Community*	*Occupation*	*Political Affiliation*
13	Mujibar Rahaman	Local leader	52	M	Graduate	Garchakraberya	Khan Muslim	Cultivation	TMC
14	Sabana Bibi	Kalicharanpur Panchayat Pradhan	40	F	VIII	Jadubarichak	Sheik Muslim	Housewife	TMC
15	Chandrasekhar Das	Kalicharanpur Panchayat Upa-pradhan	46	M	Graduate	Kalicharanpur	Paundra Kshatriya	Cultivation	TMC
16	Rabindra Das	Sanchalak	53	M	Secondary	Garchakraberya	Paundra Kshatriya	Cultivation	TMC
17	Supriya Das	Sanchalak	32	F	Graduate	7 No. Jalpai	Paundra Kshatriya	Housewife	TMC
18	Manirul Islam Khan	Sanchalak	48	M	VIII	Nagchinachar	Khan Muslim	Cultivation	TMC
19	Shivsankar Sahoo	Sanchalak	45	M	Secondary	Kalicharanpur	Paundra Kshatriya	Agricultural labour	TMC
20	Kalikrishna Pradhan	Sonachura Panchayat Pradhan	48	M	Graduate	Gangra	Mahishya	Cultivation	TMC

contd...

contd...

Sl. No.	*Name*	*Designation*	*Age*	*Sex*	*Education*	*Village*	*Community*	*Occupation*	*Political Affiliation*
21	Alpana Ari	Sonachura Panchayat Upaptrdhan	44	F	III	Soudhkhali	Paundra Kshatriya	Housewife	TMC
22	Sheik Sahabuddin	Kendumary Panchayat Pradhan	38	M	VI	Rajaramchak	Sheik Muslim	Cultivation	TMC
23	Banbihari Pal	Kendumary Panchayat Upa-pradhan	62	M	Higher Secondary	Kendumary	Mahishya	Cultivation	TMC
24	Ashok Mondal	Gokulnagar Panchayat Pradhan	40	M	Secondary	Raynagar	Paundra Kshatriya	Cultivation	TMC
25	Kanchan Das	Gokulnagar Panchayat Upa-pradhan	42	F	VIII	Simulkund	Mahishya	Ration dealer	TMC
26	Swadesh Adhikari	Sanchalak	48	M	Graduate	Gokulnagar	Paundra Kshatriya	Cultivation	TMC
27	Ahok Karan	Sanchalak	43	M	Secondary	Jambari	Paundra Kshatriya	Cultivation	TMC

contd...

contd...

Sl. No.	Name	Designation	Age	Sex	Education	Village	Community	Occupation	Political Affiliation
28	Nabadip Das Adhikari	Sanchalak	58	M	Graduate	Gokulnagar	Paundra Kshatriya	Cultivation	TMC
29	Sheik Rahaman	Satengabari Panchayat Pradhan	52	M	Secondary	Satengabari	Sheik Muslim	Cultivation	TMC
30	Sheik Kajehar	Satengabari Panchayat Upa-pradhan	48	M	VIII	Satengabari	Sheik Muslim	Cultivation	TMC
31	Satej Bhuiya	Vekutiya Panchayat Pradhan	51	M	Secondary	Vekutiya	Mahishya	Ex- Service Man	TMC
32	Chandan Das	Karmadhakha at Block level	42	M	Graduate	Kalicharanpur	Paundra Kshatriya	Service at bank	TMC
33	Bakul Dev das	Karmadhakha at Block level	60	M	Secondary	Soudhkhali	Paundra Kshatriya	Cultivation	TMC
34	Sheik Samsul	Karmadhakha at Block level	25	M	Graduate	Daudhpur	Sheik Muslim	Cultivation	TMC
35	Sanowara Bibi	Karmadhakha at Block level	28	F	Graduate	Kanchannagar	Sheik Muslim	Housewife	TMC
36	Dipak Gayen	Karmadhakha at Block level	36	M	Graduate	Mahamadpur	Mahishya	Cultivation	TMC
37	Sheik Abdus Sattar	Karmadhakha at Block level	54	M	Graduate	Samsabad	Sheik Muslim	Cultivation	TMC

Source: Field Data.

Table 3.9 Social Affiliation of Leadership of CPI(M)

Sl. No.	*Name*	*Designation*	*Age*	*Sex*	*Education*	*Village*	*Community*	*Occupation*	*Political Affili-ation*
1	Ashok Bera	Ex President of Panchayat Samity	48	M	Higher Secondary	Samsabad	Mahishya	Primary Teacher	CPI(M)
2	Iyasin ali Sah	Local Committee Secretary	58	M	Graduate	Garchakraberya	Fakir Muslim	Cultivation	CPI(M)
3	Sheik Abdul Basid	Ex Land Karmadhakha	43	M	Secondary	Osmanchak	Sheik Muslim	Cultivation	CPI(M)
4	Late Niranjan Mondal	Local Committee Secretary	50 (Age on death)	M	Secondary	Kendumary	Namasudra	Primary Teacher	CPI(M)
5	Ahok Guriya	Local Committee Secretary	52	M	Secondary	Vekutiya	Mahishya	Cultivation	CPI(M)
6	Sheik Siraj	Local Committee Member	56	M	Secondary	Kendumary	Sheik Muslim	Cultivation	CPI(M)

contd...

contd...

Sl. No.	*Name*	*Designation*	*Age*	*Sex*	*Education*	*Village*	*Community*	*Occupation*	*Political Affili-ation*
7	Sheik Abdur Rajjak	Local Committee Member	54	M	VII	Jadubarichak	Sheik Muslim	Tailor	CPI(M)
8	Debal Das	Local Committee Member	58	M	Secondary	Kalicharanpur	Paundra Kshatriya	Ex-Librarian of Government Library	CPI(M)
9	Late Himanshu Barman	Local Committee Member	53 (Age on death)	M	Graduate	Garchakraberya	Rajbanshi	Primary teacher	CPI(M)
10	Sukumar hajra	Local Committee Member	56	M	Secondary	Soudhkhali	Mahishya	Homeopath doctor	CPI(M)
11	Sheik Sahidulla	Secretary of DYFI	49	M	Graduate	Nandigram	Sheik Muslim	Service at BLR Office	CPI(M)

Source: Field Data.

of both the groups are more or less from almost similar background. The age scheme of both CPI (M) and BUPC leadership indicate the representation of middle and old aged people.

Late Nishikanta Mondal was a popular leader in the area. Initially he was a member of CPI(M) party. Later he was expelled from the party and joined TMC afterwards, when TMC came in power to Sonachura Gram Panchayat he was elected as *Pradhan* in 2008. In 2009 he was assassinated. It is alleged that he was killed by Maoists due to disagreement of opinion as reported in few newspapers.

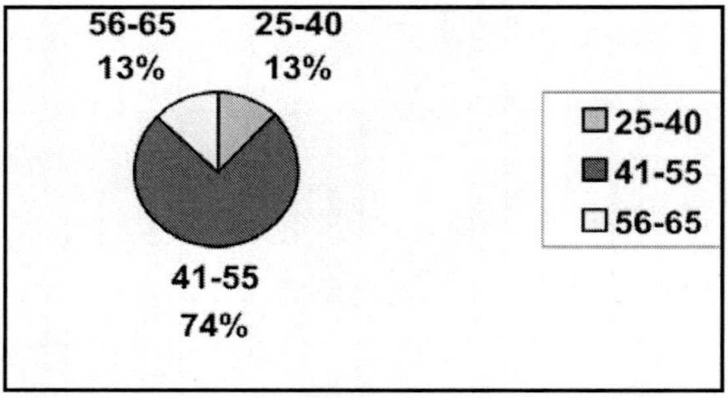

Fig. 3.4. Age and Sex-wise Distribution of BUPC Leaders

Source: Field Data.

The Fig. 3.4 has given an impression that majority of leadership of BUPC is constituted by male and with an age group of 41-55.

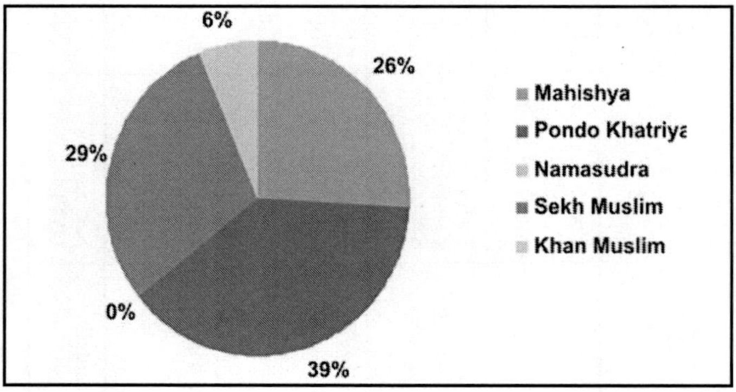

Fig. 3.5. Community-wise Distribution of BUPC Leaders

Source: Field Data.

The Fig. 3.5 shows that Paundra Kshatriya followed by Sheikh Muslim, Mahishya and Khan Muslim dominates BUPC leadership.

The Fig. 3.6 shows an idea that the leadership of BUPC is educated.

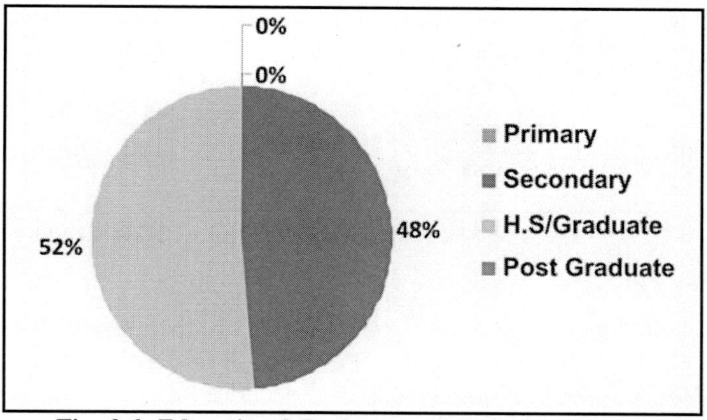

Fig. 3.6. Educational Background of BUPC Leaders

Source: Field Data.

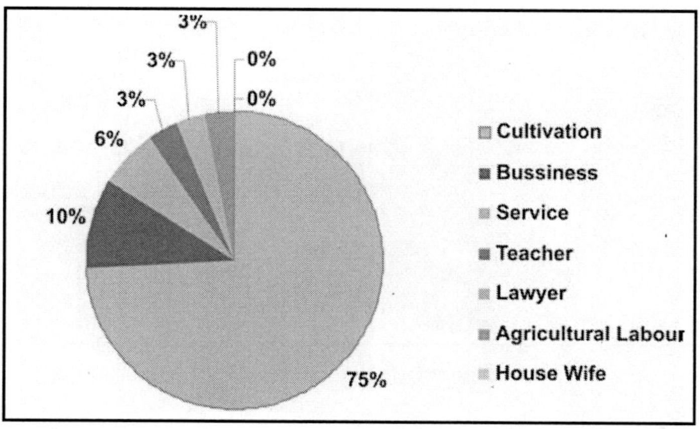

Fig. 3.7. Occupational Distribution of BUPC Leaders

Source: Field Data.

The Fig. 3.7 shows that occupation-wise BUPC leaders are distributed in wide range. In spite of that, a large section is involved in cultivation.

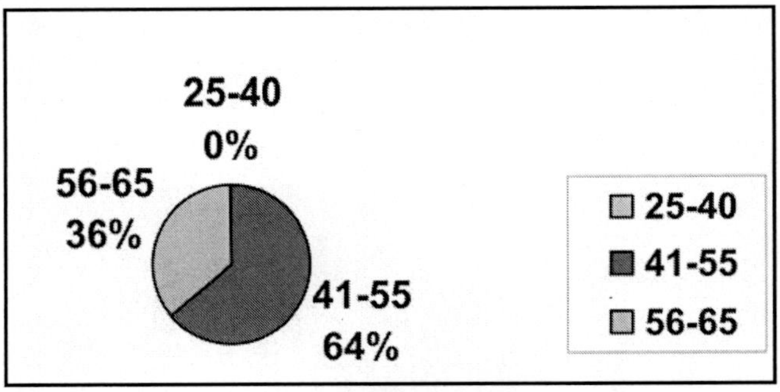

Fig. 3.8. Age and Sex-wise Distribution of CPI(M) Leaders

Source: Field Data.

The chart 3.8 has shown that majority of leadership of CPI(M) is constituted by male and with an age group of 41-55. The trend is similar to BUPC leadership.

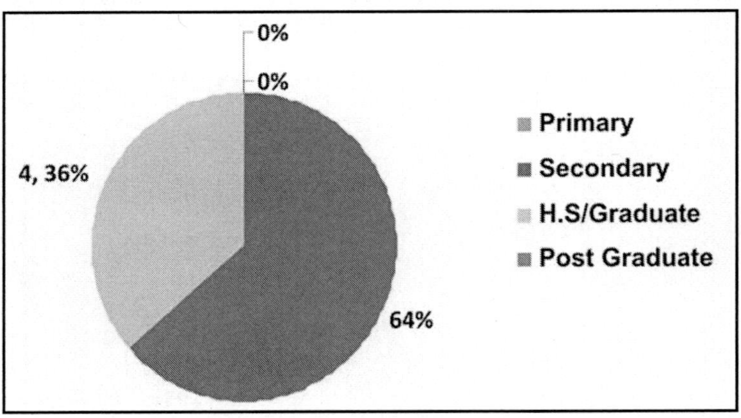

Fig. 3.9. Educational Background of CPI(M) Leaders

Source: Field Data.

The chart 3.9 depicts that majority of leaders of CPI (M) is passed secondary and higher level.

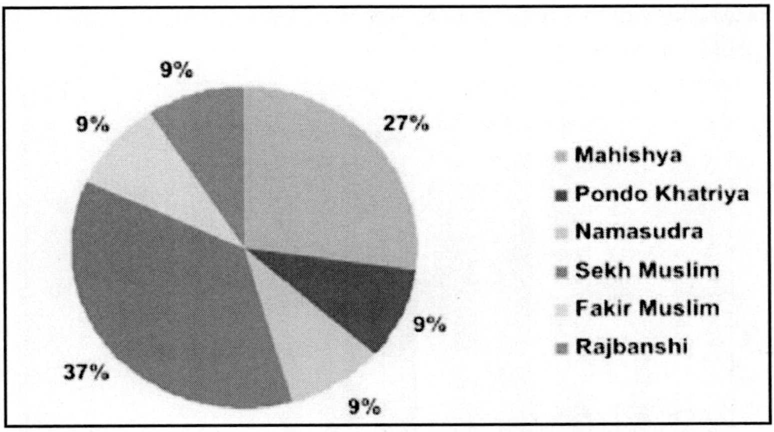

Fig. 3.10. Community-wise Distribution of CPI(M) Leaders

Source: Field Data.

The Fig. 3.10 shows that Sheikh Muslim and Mahishya mainly constitute the CPI(M) leadership.

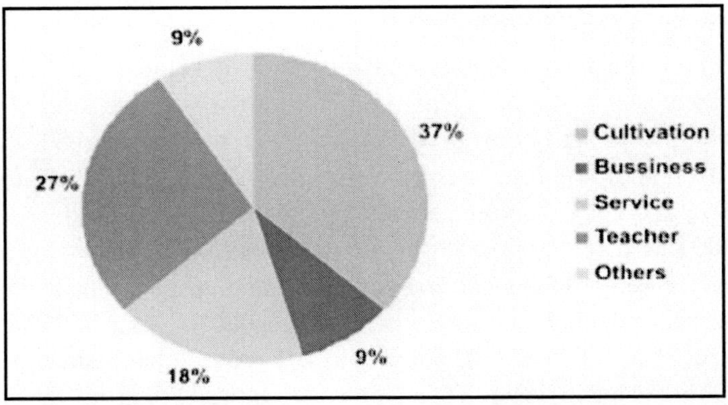

Fig. 3.11. Occupational Distribution of CPI(M) Leaders

The Fig. 3.11 shows that from occupational background main section of CPI(M) leadership is engaged in Cultivation and teaching followed by service.

The unrest uprooted the long conitued CPI(M) base and converted this left stronghold into Trinamul Congress Party (TMC) base. The following table will give an idea on shift of vote bank during electoral practice.

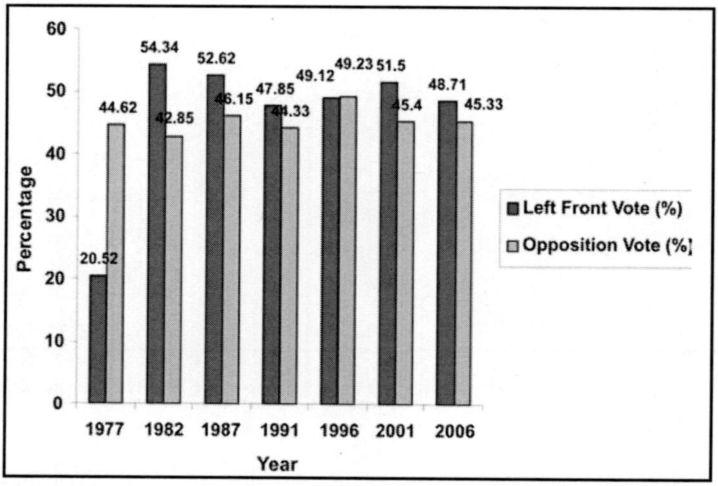

Fig. 3.12. Vote Polled in favour of Left Front and Opposition Since 1977 to 2006

Source: Election Commission of India.

From the above figures and diagrams, it could be concluded that there was a sharp increase of left vote bank started since 1982 after initiating land reforms by left front government. In 1991, there is a decline of left vote bank by 5 per cent. This is significant. Even in 1996 and 2001 the vote percentage increased and again decreased in 2006. On the contrary, the then opposition (Indian National Congress Party) vote bank was in and around 45 per cent. In 1982, there was a decline and again increased to 49 per cent in 1996. In spite of that, CPI (as Left Front partner) won election since 1982 to 2006 except in 1996. Thus, it clearly indicates that all the time opposition was a strong force in Nandigram despite the loss in electoral politics. Though the Nandigram Assembly Constituency has always shared to its main partner CPI, all the Panchayat seats remained primarily with CPI (M) as the foundation of its stronghold in the area as a whole.

Professor R.N. Maity (Former President of Nandigram Block I Panchayat Samity during 1988–1993 and CPI (M) leader) stated during an interview that in the assembly election of 2006 they won but the margin reduced to considerable extent. The voters along the coastal villages cast their votes in favour of opponent parties. He also mentioned that this is the place of Tebhaga Andolan. Bhupal Panda of CPI party the renowned Tebhaga leader was from Daudpur village and used to arrange processions along with the students and attracted general people as a magnet. The then MLA Bhupal Panda tended towards congress. Professor R.N. Maity contested against Bhupal Panda in the year 1972 and lost. The cultivable lands mainly belonged to the *jamindars* and the lands beside the river were cultivated by the sharecroppers (*bargadar*) of Muslim community. Before 1977, the sharecroppers were very influential persons because some of them used to cultivate 60 to 65 *bigha* of land and due to their influence as big sharecroppers they took over the land from the small sharecroppers. Professor Maity commented that the CPI(M) party expanded numerically but qualitatively degraded to some extent and they won three Gram Panchayats out of 10 in 2008 election.

During the interview, Professor Maity stated as it is to be understood that land (*Jami)* whether it is cultivable or otherwise ever carries or significance along with the commercial value. After the notification, a feeling was cropped up that the Left Front Government is taking away the land from the people which, in no way, can be lost as only through land one can secure own bread and butter. Large quantity of land at Sonachura is under contact cultivation (*chuktichas*) which is not registered as *barga* land. Many people are not having proper documents as *bargadar*. Compensation, if at all given, in good number cases it may be disbursed to a wrong person because of not having proper documents. He said that these are some of the delicate issues which worked against the Left Front Government.

The above discussion will be analysed in terms of two aspects. Any political movement has two dimensions one is subjective and another one is objective. The subjective dimension is the land acquisition. The proposed land acquisition was aimed for industrialisation in Nandigram. The question is what the repulsive forces are. The first one is the experience on Haldia (just opposite to Nandigram). Nandigram is indirectly benefited due to Haldia. They supply the large portion of food items required in Haldia. They have shared experience that local people were not much benefited by the industrialisation at Haldia. This parochial inclination is the first hindrance for any kind of industrialisation. The second one is the landholding character of the place. The CPI(M) leadership always pleaded that the selection mechanism was aimed to utilise the less fertile saline land. They never tried to express the truth about the landholding characters of the entire Nandigram. While, their leadership was fully aware of the situation prevailing in Gangra and Nakchirachar villages (Source: interview with Professor R.N. Maity, renowned CPI(M) leader) *bargadars* who were the unrecorded owners of 60-65 *bighas* of agricultural land, occupied this area. If only this area was initially selected, the revolt might not have been surfaced. The absentee landlords of this area welcome this move and others have no interest on the issue. BUPC leadership also extended their support on this kind of hypothetical proposition. The vast area of proposed SEZ was comprised of large number of landless people who have the vested land without lease or *patta*. They are not in a position of claiming proper compensation. Rather they would be evicted without any kind of compensation or a small amount. This kind of anxiety is due to survival threat. Presently, they are not in a prosperous livelihood condition. Even then, they are able to maintain subsistence economy. They grow rice, vegetables and procure fish from ponds, which are sufficient to sustain. Under any circumstance, they were not ready to give up the security of subsistence and invite a threatened crucial life. One informant had replied during conversation with a quotation "*Saptapurush Jethay Manush Se Mati Sonar Bara, Dainer Dai Bachiba Se Mai Amani Lakhmi Chara*" (The meaning is, the land is more precious than gold because seven generation lived here, under any circumstance this land is not sailable in spite of poverty because this land is no way less than their mother). The Muslims have the sentiment that a portion of the land is graveyard of their ancestors. Above mentioned facts were found to be the probable root causes of Nandigram unrest.

The above mentioned subjective condition was true enough to generate the unrest manifested by violence, clash etc. The subjective condition is conducive for any kind of political movement but require fulfilment of objective condition. This objective condition is the development of proper leadership who could organise the movement in every corner. The objective condition is dependent on historicity. The first portion of writing on socio-cultural dimension has discussed about the precondition for development of leadership. The next discourse on economic and political dimensions highlight the ground reality of the area to evaluate whether the other factors were responsible or not to generate the wide spread violence. The result is that the ground reality was enabled to support the aggression but not so rich to trigger the outburst of their agony without the sole cause of fear psychosis on proposed land acquisition. Any development paradigm requires to maintain impact equity in ideal condition. Industrialisation has the capacity to generate service sector for successful growth. Industrialisation is required for nation building. The problem is that a large section of local mass at the same time was depressed and suffering uncertainty due to invitation of insecurity as a consequent of land acquisition. The owners of betel leaf cultivators were opposing the land acquisition because of loss of cash crop of this area. The contractors of the fisheries were conspicuous due to good investment and uncertain return. This kind of dichotomy among beneficiaries and other sections, aggrieved of huge loss, raise questions on equity in development.

4

After-Effect of Violence and Clash

After effect is wide spread, and is multifaceted with dimensions like social, economic, political, demographic such as population structure and age-sex composition, health and nutrition and disease patterns and psychological. In this section, a micro level study in the affected villages are presented to understand the magnitude and extent of impact due to aforesaid violence at Nandigram in the recent past.

(a) Age and Sex Structure of Studied Villages

The age and sex structure refers to the manner in which proportion a population is distributed among various age and sex categories. The composition of age and sex can be considered as the building block of any population structure. The age-sex wise distribution of the population of two affected villages in Nandigram, namely, Sonachura and Garchakraberya are presented in Table 4.1. The distribution of age is divided into fourteen classes with an interval of five years i.e., 0-4, 5-9 etc. It is observed that the highest percentage of population (57.73 per cent) occurs in the age group 15-49, followed by the pre-reproductive age group (26.59 per cent), and the lowest (15.64 per cent) in the above 50 years (Table 4.2). The total population consisted of 1,316 individuals comprising 681 male and 635 females, giving an overall sex ratio of 932 females per 1,000 males.

Table 4.1. Age and Sex-wise Distribution of Sample Population in the Studied Villages.

Age Category (in years)	*Unmarried*			*Married*			*WDS*			*Total*				*G. Total*
	Male	*Female*	*Total*	*Male*	*Female*	*Total*	*Male*	*Female*	*Total*	*Male*	*Per-centage*	*Female*	*Per-centage*	
0-4	40	42	82	0	0	0	0	0	0	40	3.04	42	3.19	82 (6.23)
5-9	64	71	135	0	0	0	0	0	0	64	4.86	71	5.4	135 (10.26)
10-14	72	61	133	0	0	0	0	0	0	72	5.47	61	4.64	133 (10.11)
0-14	**176**	**174**	**350**	**0**	**0**	**0**	**0**	**0**	**0**	**176**	**13.37**	**174**	**13.23**	**350 (26.59)**
15-19	72	23	95	5	45	50	0	0	0	77	5.85	68	5.17	145 (11.10)
20-24	22	19	41	56	48	104	0	1	1	78	5.93	68	5.17	146 (11.10)
25-29	13	3	16	41	63	104	0	0	0	54	4.1	66	5.01	120 (9.11)
30-34	2	1	3	54	48	102	0	1	1	56	4.25	50	3.8	106 (8.05)
35-39	1	0	1	46	51	97	2	0	2	49	3.72	51	3.87	100 (7.59)
40-44	0	0	0	46	32	78	0	0	0	46	3.49	32	2.43	78 (5.92)
45-49	0	0	0	37	27	64	1	0	1	38	2.89	27	2.05	65 (4.94)
15-49	**110**	**46**	**156**	**285**	**314**	**599**	**3**	**2**	**5**	**398**	**30.23**	**362**	**27.5**	**760 (57.73)**
50-54	0	0	0	25	20	45	0	1	1	25	1.9	21	1.6	46 (3.57)
55-59	0	0	0	26	16	42	0	6	6	26	1.97	22	1.67	48 (3.59)
60-64	0	0	0	20	12	32	0	4	4	20	1.52	16	1.21	36 (2.73)
65 +	0	0	0	36	16	52	0	24	24	36	2.73	40	3.04	76 (5.77)
Above 50	**0**	**0**	**0**	**107**	**64**	**171**	**3**	**35**	**38**	**107**	**8.12**	**99**	**7.52**	**206 (15.61)**
TOTAL	**286**	**220**	**506**	**392**	**378**	**770**	**3**	**37**	**40**	**681**	**51.72**	**635**	**48.25**	**1316**

Source: Field Data.

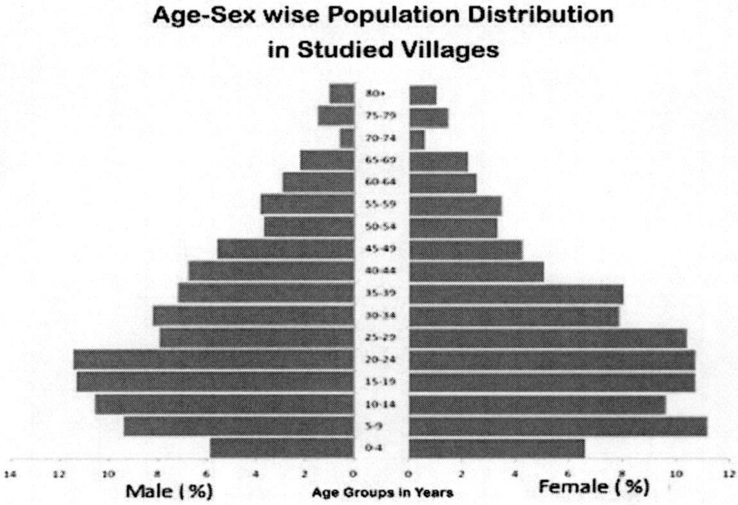

Fig. 4.1. Population Pyramid

A population pyramid has also been constructed based on the age and sex distribution of the studied population (Fig. 4.1). It may be noted that a major constriction of population pyramid at base 0-4 age group suggests that the present population of the studied villages has undergone a reduction of fertility and/or high incidence of child mortality in recent past (Fig. 4.1). The population pyramid shows the overall characteristics of a growing population, however, a threat of relatively low frequency of pre-reproductive age group (26.59 per cent) exists. It is to be pointed out that there is a considerable reduction of male members in population structure particularly in the age group 25–29 years, which is the most economically active and reproductive population may be explained due to migration or proportionally high mortality.

Marital Status

The Table 4.2 shows that the unmarried males (41.99 per cent) are considerably higher than that of female counterpart (34.65 per cent) which indicates that they are not getting suitable brides from their neighbouring areas due to recent episode as also reported by many informants during our field work. It is also interesting to mention that the women widowhood and separation is significantly higher (5.82 per cent) than that of male counterpart (0.44 per cent).

The probable reason would be that the male members in this reproductive wedlock either due to relatively high mortality of male partners or out migration (both voluntarily and forced situation) from this area which can be supported by some of the case studies presented latter in this volume.

Table 4.2. Marital Status of Studied Villages

Category	*Male*		*Female*		*Total*	
	N	*Percentage of Male*	*N*	*Percentage of Female*	*N*	*Percentage of Total population*
Unmarried	286	**41.99**	220	34.65	506	38.45
Married	392	57.56	378	59.52	770	58.51
Widow/Divorced/ Separated	3	0.44	37	**5.82**	40	3.04
Total	**681**	**99.99**	**635**	**99.99**	**1316**	**99.99**

Source: Field Data.

Household Size

It reveals from the Table 4.3 that most of the households are small in size (47.76 per cent) followed by medium size (39.17 per cent). Presently, the large extended families are found to be only 13.05 per cent. It is to be mentioned that the household size of the studied villages is ranging from single person to as many as fourteen persons.

Table 4.3. Household Size of the Studied Villages

Sl. No	*Family size*	*Sonachura*	*Garchakraberya*	*Total*
1.	1 – 4 (Small)	88	40	128 (47.76)
2.	5 – 7 (Medium)	78	27	105 (39.17)
3.	8 + (Large)	24	11	35 (13.05)
	Total	**190**	**78**	**268 (99.98)**

Source: Field Data.

Educational Status

Table 4.4 indicates that the percentage of illiterate females is almost double than the male members. However, at the primary level of education male and female members are maintaining the equal proportion about 26 per cent. It is interesting to point out that at the higher secondary level the girls outnumbered the boys in the studied villages indicating that the drop out of boys are more at this stage due to their engagement in some economic pursuits like tailoring, agricultural activities, fishing, etc.

Table 4.4. Educational Status of the Studied Villagers

Educational Status	*Male (N=625)*		*Female (N = 593)*	
	N	*Percentage*	*N*	*Percentage*
Illiterate	56	8.96	98	16.52
Literate (Can sign)	53	8.48	70	11.80
Primary	163	26.08	155	26.13
Secondary	310	49.60	241	40.64
Higher Secondary	14	2.24	19	3.20
Graduate	22	3.52	10	1.68
Post Graduate	7	1.12	0	0.00

Source: Field Data.

The trends of dropout and its magnitude have also been ascertained from the actual attendance in three primary schools and three secondary schools in the locality. Table 4.4 indicates the pattern of school enrolment in various categories during the period 2011-2012.

Table 4.5 also indicates the pattern of school dropout. The rate of dropout in case of boys is considerably higher than that of girl students. The field data corroborate this trend because of the fact that boys especially Muslim left education in search of livelihood, whereas girls continue their studies till marriage. Another important aspect is that the school going children suffered a lot due to continuous disturbance and police camps in the school establishments during the people's movement in 2007-08.

Table 4.5. School-wise Enrolment of Students (2011-2012)

Name of the School	*Class*	*Boys*					*Girls*				
		SC/ST	*OBC*	*Muslim*	*General*	*Total*	*SC/ST*	*OBC*	*Muslim*	*General*	*Total*
Sonachura Jalpai State Plan Primary School (Estd 1956)	I	2	0	0	8	10	2	0	0	7	9
	II	1	0	0	3	4	11	0	0	6	17
	III	5	0	0	3	8	4	0	0	4	8
	IV	6	0	0	1	7	4	0	0	5	9
Sonachura Board Prathamik Vidyalaya (1949)	I	17	2	0	4	23	26	5	0	8	39
	II	15	0	0	5	20	19	3	0	2	24
	III	9	2	0	4	15	15	2	0	10	27
	IV	9	2	0	7	18	14	4	0	6	24
Sonachura Khudiram Sishu Siksha Kendra (2003)	I	16	0	0	25	41	16	0	0	22	38
	II	3	0	0	5	8	6	0	0	3	9
	III	0	0	0	4	4	4	0	0	6	10

contd...

Name of the School	Class	Boys					Girls				
		SC/ST	OBC	Muslim	General	Total	SC/ST	OBC	Muslim	General	Total
Sonachura KCA Milan Vidyalaya (1964)	V	60	3	0	29	89	82	4	0	31	117
	VI	59	4	0	27	86	32	3	0	43	75
	VII	51	4	1	31	82	62	3	2	27	94
	VIII	34	4	0	26	60	55	5	2	30	92
	IX	40	6	0	26	66	44	6	0	24	74
	X	18	2	0	16	36	15	3	0	14	32
7 No Jalpai Garchakraberya Ombika Golap Smriti Vidyapith (1967)	V	2	0	56	6	64	24	0	69	37	130
	VI	3	0	53	16	72	15	0	71	26	112
	VII	4	0	29	17	50	9	0	74	46	129
	VIII	5	0	14	26	45	10	0	82	19	111
	IX	3	0	20	9	32	10	0	56	27	93
	X	2	0	17	11	30	7	0	49	8	64

contd...

contd...

contd...

Name of the School	*Class*	*Boys*					*Girls*				
		SC/ST	*OBC*	*Muslim*	*General*	*Total*	*SC/ST*	*OBC*	*Muslim*	*General*	*Total*
Kalicharanpur Dayamoyee High School (1945)	V	5	45	0	74	124	37	0	0	57	94
	VI	30	0	0	85	115	25	0	0	60	85
	VII	27	0	0	55	82	36	0	0	69	105
	VIII	29	0	0	61	90	18	0	0	32	50
	IX	32	0	0	55	87	20	0	0	52	72
	X	17	0	0	46	63	11	0	0	37	48
	XI	0	0	0	41	41	0	0	0	51	51
	XII	0	0	0	23	23	0	0	0	27	27

Source: Field Data.

(b) Health and Nutritional Profile

This section reviews the recent evidence on nutritional status of the studied villages at Nandigram Block-I. Infant and children up to age of five years constitute as much as 6.23 per cent of the studied population and from nutritional point of view constitute as vulnerable group. In the present study, the prevalence of malnutrition was judged with weight-for-height criterion recommended by WHO (2006). This indicator is used because it is independent of age for children and has an internationally accepted reference population.

The present cross-sectional study was carried out on a sample of 148 boys and 185 girls who were below five years of age. The data were collected from Sonachura and Garchakraberya villages. The mean height of below five years of boys and girls are given in Table 4.6. The mean value of height among the boys at two year age is 80.12 cm which increases to 100.89 cm at the age of five years, with an increment of 20.77 cm over a period of three years. On the other hand, among the girls at two year age is 77.38 cm which increases to 99.79 cm at the age of five years, with an increment of 22.41 cm during the same period.

The mean body weight of below five years for boys and girls are also given in Table 4.7. It reveals from the Table 4.7 that the mean body weight among the boys at two years age is 9.55 kg which increases to 13.83 kg at the age of five years with an increment of 4.28 kg over a period of three years, while among the girls at two years age is 9.13 kg which increases to 13.38 kg at the age five years with a increment of 4.25 kg over a period of three years. These children when compared with the combined Indian sample (ICMR, 1984), they are found to be marginally shorter and thinner than their counterpart.

The weight-for-height index is expressed in standard deviation units (SD) from the median of the *2006 WHO International Reference Population*. Distribution of height-for-weight indicators among the Boys (n=148) and Girls (n=185) (below five years) are presented in Table 4.8. The highest percentage of severely malnourished children (-2 SD Z-score) is observed at the age of two years and of critically malnourished (-3 SD Z-score) at the age 1.5 years of age. However, it is interesting to point out that there is a decrease in the prevalence of malnutrition among the children with increase of age.

Table 4.6. Mean Height (cm) of Boys and Girls of below Five Years Age Groups

Age (in years)	*Boys (N=148)*			*Girls (N=133)*		
	N	*Mean*	*SD*	*N*	*Mean*	*SD*
2	8	80.12	8.18	9	77.38	5.64
2.5	27	82.1	5.17	21	83.52	9.64
3	30	86.02	3.39	20	87.57	5.32
3.5	25	90.8	5.45	20	86.96	6.18
4	17	93.48	3.21	18	92.89	5.97
4.5	16	96.96	3.82	21	94.21	4.68
5	25	100.89	5.64	24	99.79	4.24

Source: Field Data.

Table 4.7. Mean Body Weight (in kg) of Boys and Girls of below Five Years Age Groups

Age (in years)	*Boys (N=148)*			*Girls (N=185)*		
	N	*Mean*	*SD*	*N*	*Mean*	*SD*
2	8	9.55	1.09	21	9.13	1.4
2.5	27	10.26	1.26	29	9.87	1.75
3	30	10.87	1.37	31	10.61	1.86
3.5	25	12.25	1.77	27	11.17	1.38
4	17	12.62	1.24	21	12.38	1.87
4.5	16	13.43	1.37	26	12.27	1.38
5	25	13.83	2.05	30	13.38	1.56

Source: Field Data.

Table 4.8. Number of Children Found Malnourished under Five Years of Age Group (N=349)

Age (in Years)	*-2 SD Z-Score*		*-3 SD Z-Score*	
	N	*%*	*N*	*%*
1	4	1.15	0	0.00
2	28	8.02	4	1.15
3	7	2.00	1	0.28
4	14	4.01	1	0.28
5	7	2.01	1	0.28
Total	**60**	**17.19**	7	**2.01**

Based on height-for-weight criterion recommended by WHO (2006).
Source: Field Data.

Out of the complete sample of 342 children below five years in the studied villages, 17.19 per cent were below Grade-2 malnourished and 2.01 per cent were below Grade-3 malnourished, which demands an urgent intervention (Fig. 4.2).

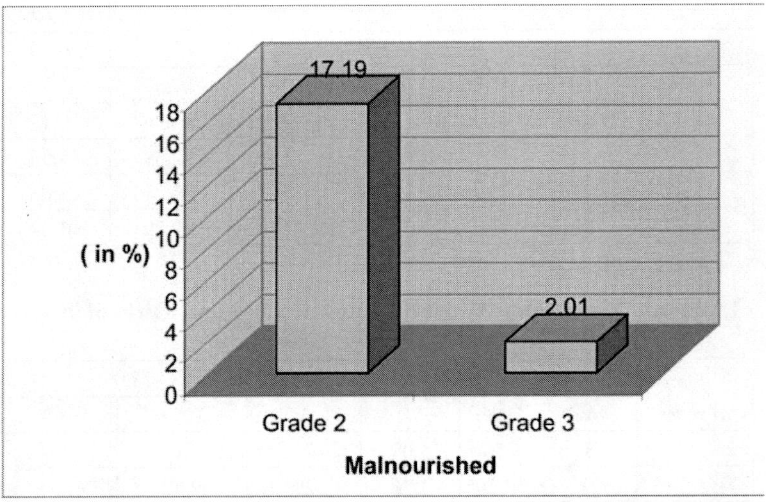

Fig. 4.2. Status of Malnutrition among the Children below Five Years

Further, it is to be mentioned that the level of under nutrition is relatively higher among the primary school going boys (29.6 per cent) and girls (29.8 per cent) of Nandigram as reported by Bose *et al.*, (2006) than that of the pre-school children below five years of age. Even, the children of the studied area at Sonachura and Garchakraberya are considered to be better off than the national average (India: 19.8 per cent in below Grade two and 6.4 per cent in below Grade-3 and West Bengal 16.9 per cent in below Grade two and 4.5 per cent in below Grade-3 (NFHS-3), suggesting implementation of adequate measure taken by the ICDS workers in child care facilities like immunisation of children with the basic vaccinations (BCG, polio, DPT, and measles) and the provision of regular health check-ups are other important components of the ICDS programme.

(c) Patterns of Morbidity and Mortality

There is enough evidences that pronounced Grade-2 or Grade-3 malnourished wasting in childhood are also associated with serious deprivations, such as ill health, susceptibility to communicable diseases, diminished learning abilities, or even higher mortality. There is a considerable circumstantial evidence to suggest a positive correlation between the status of malnutrition and the prevalence of disease pattern in the community. In the present study, the available data on the prevalence of disease pattern in Nandigram Block-I in 2008 was collected from the Block Medical Health Officer, Nandigram-I and computed which is given in Table 4.9. It is pointed out that about 90 per cent diseases are mostly confined to the communicable diseases. Of which the highest prevalence was reported from ARI (48.3 per cent) and Diarrhoeal diseases (41.16 per cent). In Nandigram the incidence of dog bite (4.62 per cent) and snake bite (0.56 per cent) are considerably high in the region.

Table 4.9. Distribution of Disease Patterns in Nandigram Block-I

Name of the Disease	*N*	*Percentage*
Diarrhoeal disease	4047	41.16
Enteric fever	184	1.87
Tuberculosis	52	0.53
Malaria	58	0.59
Acute respiratory infection (ARI)	4750	48.3
Influenza	193	1.96
Dog bite	454	4.62
Snake bite	55	0.56
Other	40	0.41
Total	9833	100.00

Source: Record of the PHC, Nandigram-I 2008.

While suffering from the curable communicable diseases as mentioned above, some non-communicable disease like Beta-Thalassaemia is also reported from Nandigram. Among 617 individuals tested for the Beta-Thalassaemia carrier, it is reported that more than six per cent from Nandigram Block-I alone (Sarkar *et al.*, 2012), which is higher than the national average of four per cent, indicating the risk of child mortality and morbidity.

With regards to morbidity of the studied area, intense protests and agitation over the issues of land acquisition have resulted in violence and loss of lives in Nandigram in recent past (2007). Besides, the consequences of current tragedy has resulted permanent and partial disability of many people in the area due to firing and serious physical injury. During our field investigation, we interviewed many such disabled as well as injured persons. Hundreds of cases of injury of different kinds and magnitudes treated in the Nandigram Block PHC, Tamluk District Hospital and two specialised hospitals in Kolkata. However, most of the victims did not reported to any medical centre in shyness or fear of social stigma particularly the women. Table 4.10 and 4.11 give an idea about the magnitude of physical injury in Nandigram on 14th March 2007.

Table 4.10. Serious Physical Injury in Nandigram on 14th March 2007

Sl. No.	*Nature of Injury*	*Male*	*Female*	*Total*
1.	Bullet injury	41	18	59
2.	Rubber bullet injury	22	15	37
3.	Fibre rod/button injury etc.	108	12	120
4.	Tear gas shell burst injury	14	12	26
5.	Bomb injury	02	-	02

Source: Samikhak Samanway, 2007.

Table 4.11. Serious Atrocities on Women on 14th and 15th March 2007

Sl. No.	*Nature of torture*	*Number*
1	Physical torture	275
2	Modesty violation	46
3	Sexual torture	17
4	Rape	11

Source: Samikhak Samanway, 2007.

Account of missing persons varies widely because many of them had left the villages due to fear or other reasons.

(d) Psycho-social Stress and Disorder of the Villagers

The study was confined to understand levels of depression and anxiety amongst the representative population who were covered under the study. Accordingly, the questionnaires on 'Anxiety', 'Depression' and 'Causes of violence' were framed and the same was administered amongst 300 individuals from two villages, namely, Sonachura and Garhchakraberia under Nandigram Block-I. Responses of the people on structured questionnaires have been analysed and discussed in following paras. The caste affiliations of the informants are Mahishya (58), Poundra Kshatriya (151), Mahato (06), Rajbanshi (14), Muslim (56) and Dhopa (15).

Depression and Anxiety

Depression or anxiety as it defined is a serious mental health concern that will touch most people's lives at some point in their lifetime (either directly or through someone close to them). Evidence from neuroscience, genetics, and clinical investigation demonstrate that depression is a disorder of the brain. Modern brain imaging technologies are revealing that in depression, neural circuits are responsible for the regulation of moods, thinking, sleep, appetite, and behaviour fail to function properly. The level of depression and anxiety depicted in Table 4.13. it reveals that the studied communities are afraid of crowd, feeling of irritation, anxiety and lack of confidence.

Incidences that caused depression/anxiety

Abuse of Power

Abuse of power, in the form of 'mal-feasance in office' or 'official misconduct', is the initiation of an unlawful act, done in an official capacity, which affects the performance of official duties. According to the survey findings, 79.45 per cent males and 55.84 per cent females of 30 to 60 years age groups of the studied population felt that witnessing sheer abuse of power caused them mental depression (Table 4.14).

Harassment

Harassment covers a wide range of offensive behaviour. It is commonly understood as behaviour intended to disturb or upset. In the legal sense, it is a kind of behaviour which is found threatening or disturbing.

Table 4.12. Educational and Occupational Background of the Respondents (Psycho-social study)

Population covered	*Sex-wise distribution*		*Education profile*							*Professional profile*						
	Male	Female	Illiterate	Literate	Primary	V-VII	IX-X	HS	Graduate	Cultivator	Business	HW	Laborer	Service	Student	Non worker
300 (20 per cent of total population)	146	154	43	43	49	100	48	9	8	71	24	135	45	10	13	2

Source: Field Data.

Table 4.13. Levels of Depression and Anxiety among the Studied Sample

Sl. No.	*Type of Anxiety*	*Response options selected by the respondent*				
		Not at all	*Occas-ionally*	*Some-times*	*Frequ-ently*	*All the time*
1	I experience shortness of breath or choking feelings (Breathing trouble)	--	---	---	146 (48.66%)	154 (51.33%)
2	I am afraid of crowds, being left alone, the dark, of strangers, or of traffic (Afraid of crowd)	--	---	---	131 (43.66%)	169 (56.33%)
3	I faint or feel like fainting (While fainting)	--	---	5 (1.66%)	65 (21.66%)	230 (76.66%)
4	I am easily irritated (Easily irritable)	--	---	13 (4.33%)	59 (19.67%)	228 (76.00%)
5	I have trouble falling or staying asleep (Uneasiness)	--	2 (0.67%)	22 (7.33%)	67 (22.33%)	209 (69.67%)
6	I am afraid of what awaits me in the future (Anxious for future)	--	---	14 (4.67%)	59 (19.67%)	227 (75.67%)
7	I worry about my health or dying (Worried about health)	--	---	7 (2.33%)	85 (28.33%)	208 (69.33%)
8	I have nightmares (Suffering from nightmare)	--	---	7 (2.33%)	85 (28.33%)	208 (69.33%)
9	I have less interest in activities that I normally enjoyed (Apathy)	--	---	7 (2.33%)	85 (28.33%)	208 (69.33%)
10	I feel good about myself (Feel good)	247 (82.33%)	53 (17.67%)	--	---	--
11	I feel I am losing control (lack of confidence)	--	---	7 (2.33%)	85 (28.33%)	208 (69.33%)

Power harassment is harassment or unwelcome attention of a political nature, often occurring in the environment of a rural set up. Sexual harassment refers to persistent and unwanted sexual advances, where the consequences of refusal are potentially very disadvantageous to the victim. During the course of study, 77.39 per cent males and 75.97 per cent females were reported that the sexual harassment as one of the cause of psychological abuse.

Humiliation

Humiliation is the abasement of pride, which creates mortification or leads to a state of being humbled or reduced to lowliness or submission. It can be brought about through bullying, intimidation, physical or mental mistreatment or trickery, or by embarrassment if a person is revealed to have committed a socially or legally unacceptable act. The study findings confirmed that about 78 per cent of the respondents (male 89.72 per cent and female 66.88 per cent) were humiliated during the period of violence.

Incivility

Incivility is a general term for social behaviour lacking in civility or good manners, on a scale from rudeness or lack of respect for elders, to damage through public drunkenness and threatening behaviour. During the period of struggle, the place has witnessed worse form of incivility. It is reported that 63.01 per cent males and 48.05 per cent females expressed about that forceful drag out of family members from their houses.

Intimidation

Intimidation is intentional behaviour "which would cause a person of ordinary sensibilities" fear of injury or harm. It's not necessary to prove that the behaviour was so violent as to cause terror or that the victim was actually frightened. The study revealed that a large number of respondents were intimidated (males 71.91 per cent and females 65.58 per cent).

Physical Abuse

Physical abuse is involving contact intended to cause feelings of intimidation, pain, injury, or other physical suffering or bodily harm. About 90 per cent of the respondents believed that physical abuse is the cause of the violence.

Table 4.14. Effects of Abuse and Impact of Proposed Project on Studied Population

Sl. No	Question	Causes of violence reported	Respondent		Non-Respondent	
			Male	Female	Male	Female
01	What are the cause of physical and psychological abuse?	i) Abuse of power	116 (79.45)	86 (55.84)	30 (20.55)	68 (44.16)
		ii)Sexual harassment	113 (77.40)	117 (75.97)	33(22.60)	37 (24.03)
		iii) Human rights violation	142 (97.26)	147 (95.45)	4 (2.74)	7 (4.55)
		iv) Humiliation by the members of the opponent group	131(89.72)	103 (66.88)	15 (10.28)	51(33.12)
		v) Acts of Incivility	92 (63.02)	74 (48.05)	54 (36.98)	80 (51.95)
		vi) Intimidation by members of opponent group	105 (71.92)	101 (65.58)	41 (28.08)	53(34.42)
		vii) Physical abuse	139 (95.20)	144 (93.50)	7 (4.80)	10 (6.50)
02	How you could be affected by proposed project?	i)My land was proposed to be acquired	146 (100.0)	154(100.0)	0	0
		ii) I would have been affected by the pollution including chemical hazards.	108 (73.97)	101(65.58)	38 (26.03)	53 (34.42)
		iii) Proposed compensation was not adequate	113(77.40)	119(77.28)	33(22.60)	35 (22.72)
		iv) Serenity of the place would have been lost.	95 (65.06)	74 (48.05)	51(34.94)	80 (51.95)
03	How you were involved in the agitation?	i) I had taken active part in demonstration	141(96.57)	148 (96.11)	5(3.43)	6 (3.89)
04	Whether proposed SEZ could have been prosperous or not?	i) Yes. There could have been prosperity and job opportunities in the area	88 (60.27)	91 (59.09)	58 (39.73)	63 (40.91)

contd...

Sl. No	Question	Causes of violence reported	Respondent		Non-Respondent	
			Male	Female	Male	Female
		ii)No, there was hardly any certainty of benefit	35(23.97)	29 (18.83)	111 (76.03)	125(81.17)
		iii) Can't say	23(15.76)	34 (22.08)	0	0
05	Do you feel this unrest will continue in near/distant future?	i)No, since the project is withdrawn	109 (74.66)	129 (83.77)	0	0
		ii)Yes, if livelihood options are not enhanced	28 (19.18)	19 (12.34)	0	0
		iii) Can't say	9 (6.16)	6 (3.89)	0	0

Source: Field Data.

It is observed from Table 4.14 that about 60 per cent of the respondent reported that there could have been some job opportunities, if SEZ would have been implemented. While about 20 per cent respondents informed that there was hardly any certainty of such benefit. About 65 per cent males and 48 per cent females were reported that serenity of the place would have been lost due to proposed project. However, at present majority of them (74.65 per cent males and 83.76 per cent females) believed that such violence would not be continued in distant future, since the proposed land acquisition has been withdrawn.

In the context of violence and political disturbance the psycho-social impact of the people of Sonachura, 7 No. Jalpai and Garcharaberia villages of Nandigram had been studied with some principles of psychology. It is reported by the informants that the act of the then local administration was not judicious by commission and omission in favour of overall consequence at that time. This evoked mental agony and fear psychosis within the people, particularly among the vulnerable section of the population like women and children. It has been reported during the field investigation that the children used to play mock fight using sticks as toy guns and marvels as bullets instead of popular game *Gulidanda*. On the other hand, women had suffered various setbacks viz. trauma, disturbance of sleep, etc.

5

Case Studies of Affected People

Case study is a method to generate first-hand qualitative information from the field in anthropological research. It provides enough scope to analyse the situation and interlink with the whole process in an analytical manner. When some cases are collected from the field and studied systematically, we may come up with some interesting facts. These facts may be corroborated or supported with the general statements as we find during initial stages of anthropological fieldwork. The case studies also provide an insight in the problems or issues in any socio-political, socio-economic or socio-cultural situation. In this context, some case studies had been collected and analysed to understand the situation at Nandigram.

It will be wise to discuss the total effect based on case study. The case studies are being classified in accordance with participation in favour of BUPC (both active and passive ways), non-participation and supporting to then ruling governments policy on land acquisition.

Case Studies Related with Active Participation in Favour of BUPC

The most acute cases are given so that anyone can understand the socio-economic background and gravity of incidence in respect of each case. Firstly, the background story is depicted in tabular forms and charts.

The following charts will elaborate and corroborate the above table. Before going to get an impression on case-to-case, there is a need to understand the socio-economic background of the affected people. The striking feature of figure 24 is 98 per cent representation of male members within the age group of 20-90 years. However,

Table 5.1. Socio-economic Background of the Affected People during Movement

Sl. No.	*Name*	*Age*	*Sex*	*Educational background*	*Village*	*Community*	*Landholding (Jointly)*	*Effect of violence*
1.	Sukhdev Das	35	M	IV	Sonachura	Mahishya	10 Cottah	Sukhdev's brother Jaidev died on 14th March, 2007 due to Police firing at Sonachura.
2.	Subodh Patra	42	M	X	Sonachura (Madhya)	Paundra Kshatriya	90 Cottah	Subodh Patra injured by rubber bullet.
3.	Sudarshan Pradhan	62	M	X	Sonachura (Madhya)	Paundra Kshatriya	50 Cottah	On 10th November, 2007 miscreants ransacked Sudarshan Pradhan's house. His two sons were abducted. At the same time his family was tortured and beaten regularly.
4.	Sujay Mondal	22	M	VIII	Sonachura (Mondal Para-Madhya)	Namasudra	80 Cottah	Sujay Mondal's education has been disturbed and discontinued.

contd...

contd...

Sl. No.	*Name*	*Age*	*Sex*	*Educational background*	*Village*	*Community*	*Landholding (Jointly)*	*Effect of violence*
5.	Shyamacharan Pal	60	M	V	Sonachura (Pal Para)	Mahishya	40 Cottah	Shyacharan Pal could not able to harvest his paddy initially. Again, he was unable to retain his 'Pan Baroch' (beetle leaf garden) due to war like situation.
6.	Arjun Pal	30	M	III	Sonachura (Pal Para)	Mahishya	20 Cottah	Arjun Pal is separated from his wife and children at present. According to his confession at present he is unable to concentrate to work due to mental agony created by misdeed of his father during 'Nandigram episode'.
7.	Subhanshusekhar Patra	36	M	VIII	Sonachura Madhya (Mondal/Patra Para)	Paundra Kshatriya	No cultivable land	Subhanshu Patra was suffered by head injury due to rubber bullet on i4th March, 2007 on police firing at Sonachura. Now he is partially physically challenged.

contd...

contd...

Sl. No.	*Name*	*Age*	*Sex*	*Educational background*	*Village*	*Community*	*Landholding (Jointly)*	*Effect of violence*
8.	Gurupada Mondal	40	M	Can sign	Sonachura Madhya (Patra/Pramanik Para)	Paundra Kshatriya	No cultivable land	He was tortured and beaten on 10th November, 2007. His backbone had been fractured and became a physically challenged person. Now he is unable to work.
9.	Paresh Patra	87	M	Illiterate	Sonachura Madhya (Patra/Pramanik Para)	Paundra Kshatriya	No cultivable land	At the time of Nandigram unrest Paresh Patra was beaten and became blind.
10.	Anjali Das	65	F	Can sign	Sonachura Sahoo Para	Mahishya	No cultivable land	Anjali Das was attacked on 14th March, 2007 and suffered by head injury. Now she is the patient of headache and vision problem.
11.	Gopal Das	35	M	Can sign	Sonachura Sahoo Para	Mahishya	No cultivable Land	Gopal Das was injured on right arm by bullet on 14th March, 2007 at Sonachura.

contd...

contd...

Sl. No.	*Name*	*Age*	*Sex*	*Educational background*	*Village*	*Community*	*Landholding (Jointly)*	*Effect of violence*
12.	Prakash Giri	58	M	IV	Sonachura Mondal Para	Mahishya	No cultivable land	On 10th November, 2007 he was injured by bullet. Now he is unable to do any work.
13.	Pratap Giri	56	M	IV	Sonachura Mondal Para	Mahishya	No cultivable land	Pratap Giri's son Rakhal Giri was died due to bullet injury on 14th March, 2007 by Police firing at Sonachura.
14.	Narugopal Gol	46	M	Can sign	Sonachura Jalpai (Gol Para)	Mahishya	No cultivable land	On 14th March, 2007 he was suffered by tear gas of Police.
15.	Gangadhar Manna	78	M	IX	Sonachura Jalpai (Dakshin Para)	Mahishya	20 Cottah	Gangadhar Manna was so shocked due to Police firing on 14th March, 2007 and suffered by brain stroke.
16.	Ratan Manna	37	M	IX	Sonachura Jalpai (Dakshin Para)	Mahishya	40 Cottah	At the time Nandigram unrest Ratan Manna was beaten on 15th March, 2007 and his left hand was injured.

contd...

contd...

Sl. No.	*Name*	*Age*	*Sex*	*Educational background*	*Village*	*Community*	*Landholding (Jointly)*	*Effect of violence*
17.	Tapan Manna	50	M	H.S. passed	Sonachura Jalpai (Dakshin Para)	Mahishya	20 Cottah	Tapan Manna was beaten and seriously injured. He is almost crippled.
18.	Jihan Khan	40	M	V	Garchakraberya (Khan Pally)	Khan Muslim	40 Cottah	Jihan Khan suffered injury on back and pelvis on 8th March, 2007.
19.	Sheik Abed	55	M	Illiterate	Garchakraberya (Paschim Para)	Sheik Muslim	No cultivable land	One police case is registered against him accusing snatching the gold shop and medicine shop on 10th November, 2010.
20.	Sheik Amjed	32	M	Illiterate	Garchakraberya (Paschim Para)	Sheik Muslim	No cultivable land	One police case is registered against him accusing snatching the gold shop and medicine shop on 10th November, 2010.

Source: Field Data.

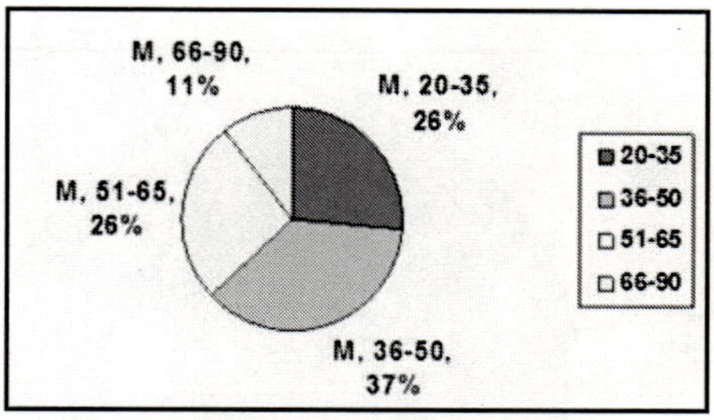

Fig. 5.1. Age & Sex Wise Affected People

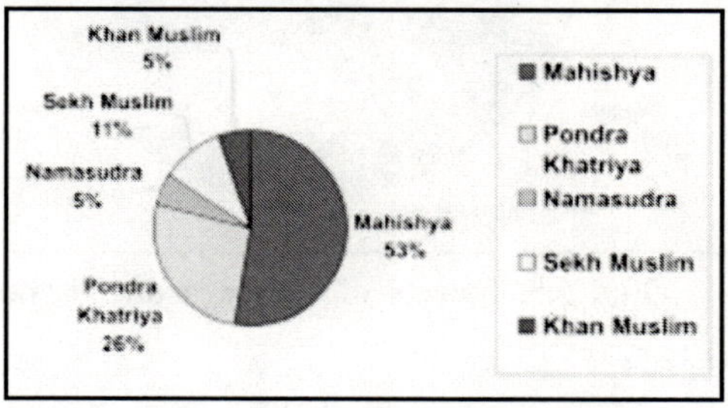

Fig. 5.2. Community Wise Distribution of Affected People

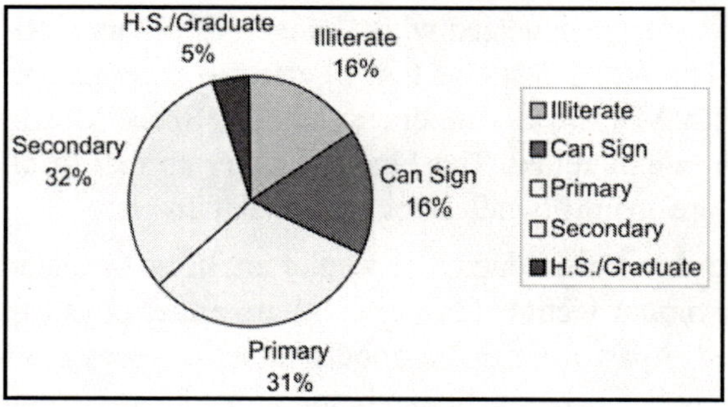

Fig. 5.3. Educational Background of Affected People

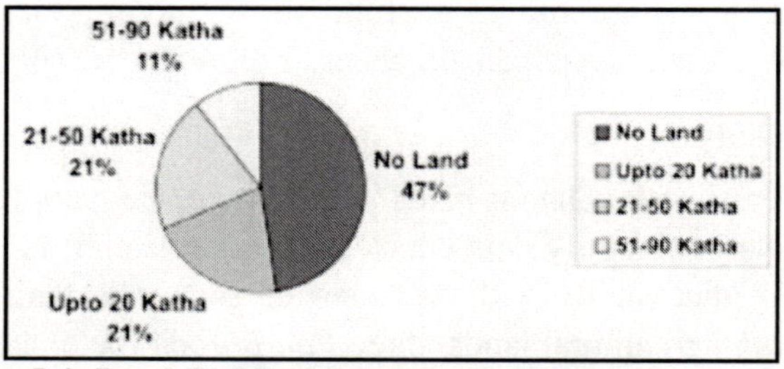

Fig. 5.4. Land Holding Pattern of Affected People

Source: Field Data.

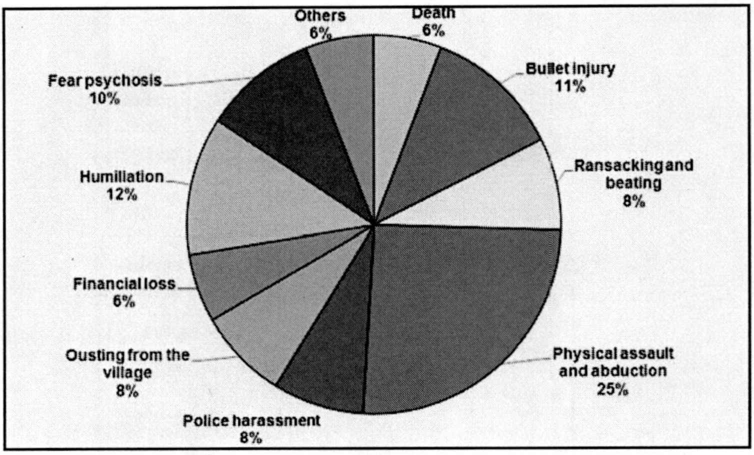

Fig. 5.5. Effects of Violence on Victims

Source: Field Data.

the maximum participants recorded as case studies are within the age group of 36–65 years. The Fig. 5.3 shows that in 68 per cent-cases people are educated with a range of Primary to H.S./Graduate Level. The Fig. 5.2 shows that in affected cases people are mostly belong to Mahishya, Paundra Kshatriya, Sheik Muslim and Khan Muslim communities. The Fig. 5.4 indicates that in affected cases people are mainly landless or marginal farmers.

The above mentioned case studies implicate the nature of victims. The maximum victims (24 per cent) are either physically assault or abducted. Apart from that a good number of cases are found which indicate police harassment and humiliation due to participation in the demonstrations. In the above chart different categories of victims are shown where others include those having family relation disturbed, educational disturbances, missing, etc. (Fig. 5.5).

Case Study No. 1

Sudarshan Pradhan is of 62 years of age. He belongs to Paundra Kshatriya community. His educational qualification is IXth passed. His is a joint family with two married sons. His family possesses 50 *Cottah* agricultural lands, three *Pan Boroch* (betel leaf vines). He alleged that on 10th November, 2007 the supporters of the then ruling party ransacked his house. Both the sons were abducted and kept in captivity at Serkhanchak village of Kjejuri block. On 11th November, 2007, after regaining the control on entire village, they were ordered to obey the dictum of the then political masters of the village. At the same time, they had been tortured and beaten regularly.

Case Study No. 2

Shymacharan Pal is of 60 years of age and class V educated. He belongs to Mahishya community. He is V class passed. He has four married daughters and one married son. This is a joint family. His son Krishnaprasad Pal is presently working at Kolkata. He earns Rs 3000 per month. His family possesses 40 *cottah* agricultural lands. At the time of Nandigram unrest, Shyamacharan Pal went to his in-laws house due to enormous pressure on him. He could not harvest his paddy initially. After Panchayat election in May, 2008 he was somehow able to manage the harvested paddy confessing his 'misdeed' due to being a sympathiser of BUPC. Again, he was unable to retain his '*Pan Baroj*' (betel leaf plantation) due to war like situation.

Case Study No. 3

Gurupada Mondal is of 40 years of age. He belongs to Paundra Kshatriya community. He can only sign. He has three married brothers and three married sisters. His all brothers are engaged as labourers at Sagar Island of South 24-Parganas district of West Bengal. His family is not having agricultural land. Gurupada Mondal was a supporter of local political party. He is a BPL card holder. Fifteen years back, he constructed his own house under the Govt. scheme 'Indira Awaas Yojona'. By occupation, he was agricultural labourer. At the time of Nandigram unrest, he declined to give his land and joined to BUPC. As a result, he was tortured and beaten on 10th November, 2007. His backbone had been fractured and he became a physically challenged person. Now he is unable to work.

Case Study No. 4

Anjali Das is of 65 years of age. She belongs to Mahishya community. She can only sign. Her husband's name is Mritunjay Das. She has three married sons. Her family is not having any cultivable land. According to her statement, she was attacked on

14th March, 2007. After head injury, she was unconscious on ground. Later she was admitted to Tamluk hospital and then referred to P.G. hospital at Kolkata. She was treated there for two months. Still she is suffering from headache and vision problem.

Case Study No. 5

Pulin Bihari Jana is 65 years of age and belongs to Mahishya community of Sonachura Maddhya (Patra/Pramanik Para) village. He has passed VII standard. His family has 30 *Cottah* agricultural lands. Pulin Bihari Jana applied for old age pension two/three years ago but still he has not been granted. He is BPL cardholder. He was granted Rs15000/- under 'Indira Awaas Yojana'10 years back. His family was passive towards the unrest led by BPUC. Despite that, they took part in movement by compulsion.

Case Study No. 6

Samir Jana is of 29 years age and belongs to Mahishya community of Sonachura Purba Pally (Jana Para) village. He is class H.S. passed. His family has 30 *cottah* agricultural lands. Samir Jana participated in procession and beaten on 10th November, 2007.

Case Study No. 7

Bantulkrishna Kajli is of 67 years of age and belongs to Mahishya community of Sonachura (Sahoo Para) village. He has passed upto VI standard. His family has 30 *cottah* agricultural lands. Batulkrishna Kajli's family members like Gobinda Kajli, Uttam Kajli, Madhuri Kajli and Pratima Bera took part in the procession organised by BUPC on 10th November, 2007. It is alleged that they were beaten and abducted by the then ruling party supporters. Later, they were used as frontier soldiers (*dhal*) while entering inside Nandigram.

Case Study Study No. 8

A Muslim family from lower income group from the Garchakraberya village had been tortured by a group of people during the time of the conflict. Four of its members were affected directly. One member of this family experienced severe fracture on the head and his arms were broken due to the physical assault.

Table 5.2. Socio-economic Background of the People Passively Participated in the Movement

SL. No.	Name	Age	Sex	Educational background	Village	Community	Landholding (Jointly)	Effect of violence
1.	Parameswar Maity	60	M	II	Sonachura	Mahishya	20 Cottah	He was suffered due to tear gas while took part in procession.
2.	Chabirani Maity	45	F	Illiterate	Sonachura	Mahishya	20 Cottah	She was suffered due to tear gas while took part in procession.
3.	Jagadish Chandra Maity	40	M	VII	Sonachura	Mahishya	20 Cottah	Jagadish Chandra Maity and his wife were suffered due to tear gas while took part in procession.
4.	Tulsi Mondal	30	F	Can sign	Sonachura	Paundra Kshatriya	5 Cottah	He was beaten.
5.	Dulal Mondal	25	M	Illiterate	Sonachura (Bisweswari Pally-North)	Paundra Kshatriya	5 Cottah	He was beaten.

contd...

contd...

SL. No.	*Name*	*Age*	*Sex*	*Educational background*	*Village*	*Community*	*Landholding (Jointly)*	*Effect of violence*
6.	Ashok Mondal	52	M	Graduate	Sonachura (Mondal Para-Madhya)	Namasudra	160 Cottah	On 14th March, 2007 and 7th August, 2008 miscreants ransacked Ashok Mondal's house.
7.	Sabuj Patra	39	M	VII	Sonachura Maddhya (Patra/Pramanik Para)	Paundra Kshatriya	Landless	He along with his family was in camp at Nandigram block office at the time Nandigram unrest (from May, 2008 to November, 2010)
8.	Pulin bihari Jana	65	M	VII	Sonachura Purba Pally (Jana Para)	Mahishya	30 Cottah	He took part in movement by compulsion and threatened.
9.	Samir Jana	29	M	H.S.	Sonachura Purba Pally (Jana Para)	Mahishya	30 Cottah	He participated in the movement and was beaten.

contd...

contd...

SL. No.	Name	Age	Sex	Educational background	Village	Community	Landholding (Jointly)	Effect of violence
10.	Prabir Jana	24	M	IX	Sonachura Purba Pally (Jana Para)	Mahishya	30 Cottah	He was captured and beaten.
11.	Bantulkrishna Kajli	67	M	VI	Sonachura Sahoo Para (Kajli-Eastside)	Mahishya	30 Cottah	He and his family members took part in the procession on 10th November, 2007. They were beaten and abducted. Later, they were used as frontier soldiers ('Dhal') to enter inside of Nandigram.
12.	Monoranjan Jana	67	M	Secondary	Sonachura Mondal Para	Mahishya	50 Cottah	He participated in the movement and was beaten.
13.	Prankrishna Sanki	62	M	IV	Sonachura Jalpai	Mahishya	50 Cottah	He participated in the movement and was threatened.

contd...

contd...

SL. No.	Name	Age	Sex	Educational background	Village	Community	Landholding (Jointly)	Effect of violence
14	Sheik Ajit	67	M	Illiterate	Garchakraberya (Paschim Para)	Sheik Muslim	Landless	He was evicted from the village. He could enter to his own house at the time of Panchayet Election on May, 2008.
15	Sheik Abdul Jabbar	53	M	IV	Garchakraberya (Paschim Para)	Sheik Muslim	70 Cottah	He was evicted from the village. He could enter to his own house at the time of Panchayet Election on May, 2008.
16	Sheik Ajijul	40	M	Illiterate	Garchakraberya (Paschim Para)	Sheik Muslim	Landless	He participated in the movement and was threatened.

Source: Field Data.

Sheikh Hatem, aged 60 years, used to maintain his livelihood by working as daily labourer and agricultural labourer. He got four decimal *patta* lands from the panchayat. He passed out class-IV and the supporter of the Left Front government for a long considerable period of time. They heard the news that the government had decided to take over the land from some of the villages of the Nandigram Block-I including his village of Garchakraberya to set up a chemical hub there. Immediately, he himself and other family members shifted their support in favour of Bhumi Ucched Pratirodh Committee. His opinion and opinion of his other family members reveal that torture came as a result of shifting of support from their previous political party to a new political party. His wife Sahahadi Bibi, a literate person of 55 years of age, was beaten by the followers of the Left Front supporters. She was dragged by legs and thrown away into the nearby pond.

Sheikh Hatem is the father of three sons and three daughters. Among them only one son and one daughter were unmarried and others were married. That unmarried daughter named as Ruksana Khatun, aged 17 years, dropped out her study after fourth standard and now helping her mother in domestic activities. The unmarried son named as Seikh Yasin, aged 15 years, dropped out his study after passing fifth standard and left the village for tailoring works somewhere in the state of Bihar. He left the village after the agitation to earn his livelihood which is a significant trend at Nandigram, but it is not clear whether he left the village due to the agitation or not.

Case Study No. 9

Sheikh Sahalam, aged 27 years, left out his study after passing fourth standard was present at the time of agitation at Garchakraberya village. He was also attacked by the followers of the left front government with rods, sticks, etc. within the jurisdiction of their homestead area and he was injured having fractured his left hand and legs. After that incident, he left the village for Bilaspur of Chhattisgarh state for tailoring work for his livelihood. There, he gets monthly remuneration of Rs 2,000 and he sends some amount to his family or when he visits his village after five or six months he brings a lump sum amount for his family. His wife named as Rohima Bibi stays at Garchakraberya village. She is only 23 years old and dropped out of her study after passing fourth standard. Her birth place is Kendamari village under Nandigram Block-I. She experienced a still birth one-and-a-half year before. Habibur Rahaman, another son of Sheikh Hatem, aged 22 years dropped out his study after passing fifth standard married to Sahara Bibi of Garchakraberya village. She is the mother's sister's daughter of her husband. He also left the village after marriage in Kolkata and gets Rs 2,000 per month as remuneration for his service and some amount is sent to his wife to run their livelihood.

Sheikh Hatem's two daughters married outside the village, one at Haldia and another in Delhi. Another daughter, mentally ill, married to Ajanur Mallick of the same village and left for Delhi in search of work. Now, they are working as ragpickers to earn a living.

Other family members also suffered mentally as well as socially due to some social restrictions imposed by them. They could not yet forget it and sometimes are affected by fear psychosis still now.

Case Study No. 10

Sheikh Khokon, brother of Sheikh Hatem was also tortured at that time. He married to the Ansura Bibi of Garchakraberya village. She is her husband's maternal uncle's daughter. Sheikh Khokon is the father of one unmarried son and three unmarried daughters. One of his daughters died after three months of delivery due to fever and cough and cold in the year 2007 which is the most turbulent period. He said that he was very busy to protect themselves from threats of life and hardly got sufficient time to provide attention and care to the new born baby which was absolutely necessary. Ansura Bibi mentally suffered a lot at that time and somehow managed her family very painfully. Her other two daughters and one son are studying in different classes at local schools at present but their education suffered a lot at that time which was expressed by both of them. Her husband belongs to below poverty line category and got four-decimal *patta* land from the local panchayat of the previous government. The family was also the staunch follower of the Left Front Government for a long period of time, and gradually shifted towards the opposition which may be one of the causes for being tortured by their earlier counterparts. There are many such types of cases observed during field investigation wherein total family members were shattered due to the conflicts and confrontations pivoted around the decision of land acquisition. Now, they are without fear to share their experiences with the outsiders like ourselves. Whole family members irrespective of age and sex as well as of caste and religion talked freely which was noted during our investigation. These families are gradually overcoming the situation with the support of the local village society. A kind of social transformation has taken place and each one is very alert in this respect during any kind of social interaction. Their inherent agony has not totally washed away and very rarely they interact socially in any kind of ceremonial occasion of the staunch followers of the then ruling party. Many of them left the villages to hide out in an unknown place, or in a relative's house outside Nandigram with a fear of revengeful retaliation. Some of them have compromised with the present local leaderships and staying at the village itself. The fear psychosis is still rampant within the village society among two distinct groups of people.

Case Study No. 11

Nikhileswar Giri, aged 58 years, passed fifth standard and married to SibaniGiri of Soudkhali village of Nandigram Block-I. They belong to Mahishya caste and Sandilya clan. Late UpendraGiri, the grand-father of NikhileshGiri came to Sonachura-Jalpai, some 100 years ago, from Takapura village under Nandigram police station. Nikhileswar said that southern part of Sonachura-jalpai was situated on the banks of Hugli River and covered with dense forests at the time of his grandfather. Late Upendranath came along with a m*ahajan* (landlords), late Bhupal Chandra Bhuiya of the same village. Primarily there was two *bighas* of cultivable *barga* land with late BagambarGiri, the father of NikhileshGiri. Out of these two *bigha*s of land, an internal settlement had been done between the *mahajan* and the share-cropper *(bargadar)* late BagambarGiri in which later on late Bagambar purchased one *bigha* paying Rs 35,000 and the *Dalilnama* was prepared in the names of his four sons with 21 decimal land each. In this process, Barga-land had been transformed into the personal ownership through purchase leaving aside half of the *barga* land unaltered.

Case Study No. 12

Cases had been framed against NikhileshGiri for violation of law at the time of 14th March, 2007 including a murder case against him. He used to attend the cases at Haldia court. The cases had also been framed against some persons who were not present in the village at that time as stated by the villagers. NikhileswarGiri was a follower of Congress party for two decades. Later on he became the supporter of the Left Front. He supported the Bhumi Uchchhed Pratirodh Committee which later on came under the TMC party which is ruling the West Bengal State at present. When asked about his feeling at present after change of the government he replied "*Ekhan Shanti Shanti" (Now it is complete peace). "BamfronterAtyacharTheke Shanti Mileche*" (releaved from the torture of the Left Front government). He also stated that during one and a half year after the incident on 14th March, 2007, the followers of Left Front marked and attacked the houses and family members of TMC supporters with arms and weapons. A local Left Front leader was behind the atrocities at Sonachura-Jalpai. When TMC came to power, that leader left the village but his brothers are still living there. The local people are cooperating with them. It is heard that he is residing at Haldia town at present and his wife got a service at Haldia Hospital, as nurse.

Case Study No. 13

NikhileswarGiri's wife gave birth to two sons and three daughters. Her first two daughters died very early. One died at the age of six months due to severe diarrhoea, locally called *taksin.* Madan Giri, the son of NikhileshGiri, also faced the atrocities of the Left Front. He was 35 years of age, studied up to fifth-standard, and was mainly engaged as agricultural labourer and also used to cultivate his father's cultivable land. He joined other villagers during the peace procession *(Santimichhil)* on 10th November, 2007, when he was dragged and cornered in a kind of prisoners' camp behind a school. In that camp, Madan Giri and many others got rough treatment from the followers of Left Front. It was stated by many villagers that the BUPC supporters held in that camp was regularly tortured. Those who could not bear the beatings and fainted, was given medical treatment. It continued for a long time. Madan Giri still can remember that the incidence of getting tortured by wooden sticks, iron rod, etc. and that had badly damaged his lower back portion. Though recovered but still was not in a position to work properly. His wife Aparna Giri participated in the procession of 14th March, 2007, when 14 peoples died in police firing. Madan Giri expressed his anger that the incidence has not only damaged his economically and physically but also in terms of education of his son and daughter badly. Although passed out from their respective classes but their results were not up to the mark.

Case Study No. 14

Madan Giri's father, NikhileswarGiri, mentioned that another son KartikGiri was approaching the marriageable age at that time. But he had to leave the village intermittently due to fear of atrocities. After the change of power from Left Front to TMC in the year 2011, a relatively peaceful atmosphere has returned in the village life and Kartik was able to stay and perform his daily activities and managed to get married. He had an eleven months old son during the present field investigation.

Such case studies have revealed many abnormalities in the sphere of economy, education as well as social disconnection within the village society, due to conflicts and confrontations related to the decision of land acquisition of the then Left Front Government. Another case study is described here to mark the atrocities by the followers of Left Front on the family members of Suresh Chandra Gol, Mahishya by caste, aged 85 years. Now, he is a widower and residing with his elder son at Golpara of Sonachura village. The migration history of this family is traced back to the grand-father of Suresh Chandra Gol, who came here from Ranichak. Sureshbabu told that he had seen the Tebhaga movement at this village and also experienced the *Jail-bharo* agitation *(Andolan)* at the time of British

India. But the Nandigram agitation is the worst of its kind and most terrific in comparison to the above mentioned movements.

Case Study No. 15

Suresh Gol and his family members were supporters of the Congress party, but his elder brother Late Paresh Chandra Gol was in favour in the CPI (M) party. This may be one of the reasons to target Sureshbabu's family members. His elder son DebabarataGol, aged 58 years, previously suffered a mild stroke and one side of his body is affected. He was not in a position to do heavy work due to his disability. A group of people backed by Left Front supporters entered his house at night when he was watching television. People entered inside the room with arms and daggers in their hands and ransacked household items including television, dish antenna, furniture, etc. Debabratababu was terribly shocked and trembling during the incident which has made a permanent impact in his psyche. The family members told that NabaSamanta, local influential CPI(M) leader and his *bahini* (followers), played behind the scene, which was also mentioned earlier in another case study. On 14th March, 2007, DebabrataGol's son SubolGol and his wife NamitaGol attended the procession towards Bhangabera Bridge and police fired bullets and tear gas in that. The tear gas affected eyes of many villagers who attended the procession. SubolGol and NamitaGol were also affected by tear gas. NamitaGol has developed night blindness since then. The schools were closed at that time for one month. Subol's two sons could not attend the schools at that time; and passed out with average marks.

Case Study No. 16

Shri Satish Jana, aged about 70 years used to live in Sonachura village since his birth as second child of his parent. He lost his wife who died due to heart disease some 12 years back. He had two *bigha*s of fertile land where cultivation was being done by his son Shri Sukumar Jana, 47 years and his daughter-in- law Smt. Supriya Jana, aged 35 years. His life was otherwise happy with grandson, granddaughter and her husband Madan Mohan Mondal. Mondal life, however, was shattered as his daughter-in-law Supriya was injured due to police action when she took active part in protest rally. However, she received payment of Rupees five lakh from the Government of West Bengal later. Undoubtedly, entire family of Shri Jana appeared to be shaken and under deep state of depression.

Case Study No. 17

Smt. Paru Rani Jana aged 60 years is residing in Sonachura village. It is reported that when she went to attend a meeting on 10th November, 2007, few persons fired indiscriminately at that crowd resulting loss of one her fingers in right hand permanently. She reported that no compensation has yet been paid to her for this 'loss'.

Case Study No. 18

Smt. BinduBala Das aged 70 years, residing in Berapara village, had lost her 23 year old son Jaideb Das on 14th March, 2007. Late Jaideb became victim of police firing. A sum of rupees five lakh was given to that family as compensation. It has been told by Jaydev's nephew that, Abhijit who was only four years old started suffering from trauma due to his uncle's unfortunate death.

Case Study No. 19

Sri Sanjay Mondal, 27 years old, belongs to Sonachura village. At the time of unrest at Nandigram his wife was 23 years old and in the early stage of pregnancy. His wife was alone at that time. As a consequence of disturbance and loneliness she suffered mental stress and strain. Then a male child was born in November, 2007. As the child grew up they observed that the he speaks very little and has developed fear psychosis about unknown male members around his little world.

Case Study No. 20

Shri Shankar Prasad Das aged about 28 years, from Sonachura village pointed out that his father Shri Narayan Das aged about 65 years has been untraceable since 14th March, 2007. Shri Das pointed out that his old mother Smt. SajalBala Das refused to believe that her husband perhaps will never return. His entire family seems to be in a sombre mood.

Emerging Issues from Above Case Studies

These case studies show that many families have suffered a lot almost in every sphere of life experiencing social humiliation, mental unrest, physical assault, economic disorganisation and political upheaval. Such types of incidents, as depicted by many family members of that locality, have created a deep rooted impact on their minds and bodies. Still, they are gradually overcoming such ferocious incidents because of the changing socio-political situation at both local and state level.

The unfortunate incidence, seemed to have occurred due mainly to the misconception and lack of information about the socio-economic and socio-political situation of the people of the area. Land (*jami*) is very precious to them both for the land-owners and land-less peoples. Land-less people indirectly depend on other's land for a living as agricultural labourers and contract cultivator. Land is to be seen not only as a commercial point of view but also from social point of view as it has immense social significance.

Report suggests that children, who are affected by interpersonal and intrapersonal violence, including psychological aggression and verbal aggression, may exhibit a range of serious disorders, including chronic depression, anxiety, posttraumatic stress disorder, dissociation and anger.

During the survey it was witnessed that all vulnerable groups of people namely, children, women and elders have suffered a lot psychologically during the time of struggle in Nandigram. The incoherent behaviour patterns amongst a group of population, admittedly, is still in existence as routine behaviour and cohesive interpersonal relationship which is natural in rural set up in our country is missing in Nandigram and more precisely in Garhchakraberia and Sonachura villages.

6

Peoples' Perception and Related Issues on Land Acquisition

This section highlights the related issues suggested by the members of the sampled household, which may address how this kind of conflict could be averted. Peoples' view on relevant problem and long standing issues are discussed below:

I. Predictable Impact Areas

(a) Land Ownership

A large section of landless villagers, i.e., *bargadars* and contract cultivators, etc. (48.36 per cent) are deprived of any authenticated title of the land (*patta)* on which they are dependent. Even after the large quantities of vested lands have been distributed to landless farmers through land reforms, but most of them did not get the title right of these lands. They want immediate ownership right of their lands by virtue of *patta* (lease) or *rayat* (ownership) so that they will be eligible for any compensation and rehabilitation package.

(b) Infrastructure Development

The local people have grievances on developmental aspects. These aspects are the following: (i) Two bridges–Rasulpur Bridge, Haldia-Nandigram Bridge are long pending demand of the local people. (ii) Rural electrification is the aspiration of rural people, which they expect from the government. By providing infrastructural facility to rural people, the distance between urban and rural could have been minimised.

(c) Scope of Industrialisation on Local Resource Bases

The process of industrialisation was not given adequate importance which could benefit the local people. The potentiality of agro based industry, considering the growing agricultural surplus, has not been explored meaningfully so that people get confidence on the entire process of industrialisation. There are ample scopes in Nandigram on this issue. For instance, the vegetables, betel leaves cultivation and inland fisheries may generate the agro based commercial enterprises which render support to the farmers to sell their products at local level and understanding the demand of market. Thus, the people wanted agro based industrial development in this area. The three demands are linked with the problem of peasants that they want to resolve immediately by getting: (a) easy agricultural loan, (b) high grade seeds, and (c) special package for marginal farmers. Besides, they also demand to provide irrigational facilities to their agricultural lands. Still they hope for bright prospect of agriculture in this area.

(d) Administrative Transparency and Confidence Building

The peoples' demand was a good and wide compensation package to include even the landless and marginal land holders. Even land owners were in dilemma on the size of compensation package and after hearing about land acquisition they started to deposit the due land taxes immediately to avoid the delay in getting compensation.

There was lack of transparency; the people did not know the agreement between corporate sector and the government. The allegation was that the government was not loyal to the people, rather loyal to corporate sector. In case of land acquisition people were suspicious on the integrity of government on that issue. According to peoples' version the local peoples' representatives have no credential to prove pro-people move on any matter. This is why they have lost confidence. Primarily these people were not adverse to industrialisation. They wanted a true public hearing on this issue where the representatives of government could converse directly to the people.

If we look at the views and demands of the people of Nandigram, the real issues have emerged. Before land acquisition, there is need of proper evaluation of long pending demands of the people.

II. Aspirations on Industrial Project

It is obvious that in a traditional agrarian economy with limited experience of agricultural development and a large number of landless sharecroppers and agricultural labourers, search for alternative livelihood options become stronger and critical. On the other hand, the contrasting level of development through modern industrialisation process in the adjacent Haldia enhanced the aspirations on industrialisation among the people of Nandigram. But at the same time, the quantum of benefits received by the local people through compensation and direct or indirect employment generated had remained far below their expectations both in case of Haldia Petrochemicals and Jelingham project.

In course of the present study, somewhat confusing and contradictory opinions were presented by the people of Nandigram, five years after their full throttled resistance against the land acquisition and industrialisation. A majority of the people expressed that primarily they were never against industrialisation rather their level of aspiration was high. On the other hand, the imposed acquisition and its fall out in Singur fuelled the fear psychosis and disbelief in the government's measures and policy in this regard. Further, the absence of official format in details of the proposed project and somewhat whimsical announcement of the same created a complex situation within a very short period of time within the already existing background of confusion.

III. Interest Group and Driving Forces

In fact, there were differences in opinion and aspiration among at least three distinct interest groups within the proposed project area. The first being the erstwhile landlords or *mahajans* who have lost their holding through Land Ceiling Act, and had negative or no interest on the riverside land given to *bargadars* or *pattadars*. Secondly, the rich agriculturists having larger landholdings and good returns thereof can be seen as an aspirant group for highest possible compensation for land. The legal and unofficial *bargadars*, though suffering as a marginalised holder, were found split in two groups.

The legal one could visualise direct compensation whereas the other one found it as the last nail on the coffin.

The most predictable opposition group were the recently flourishing brackish water prawn fishery owners, who have invested huge money and taken a high risk on a land not or illegally owned/ leased in by them.This group, having illegally converted the character of the land either from community fishing land or cultivable land to brackish water fisheries, were neither in a position to claim compensation nor could return the land at its original state to its actual owners. On other hand, the illegal transeferers of *barga* or *patta* land also found it improbable to get compensation if any instead of the amount of investment, extent of business risk and the increasing heavy returns.

At the onset of Nandigram unrest, the area of protest was limited to Kalicharanpur Gram Panchayat to Sonachura Gram Panchayat in anticipation of the land acquisition. The movement in January, 2007 was the culmination of protest in this area only. After 14th March, 2007 (the incidence of police firing) the shift of interest groups were noticed. Especially those who were the support base of the then ruling party turned into opposite camp in apprehension of acquisition of their inland fisheries which was one important economic entrepreneurship in this locality. Hence, afterwards the protest was further strengthening by involvement of more people. Again, after 10th November incidence a large section of the entire area came in support of BUPC due to propaganda that in future the other area comprising few portions of Nandigram Block-II and Khejuri Block II would come under the proposed chemical hub. The interest groups were being continuously lengthened and consolidated such type of violence as observed from the chronology of events. The village society had turned into two distinct divisions, socially and politically, due to hidden anxiety of proposed acquisition, culminating into violent clash. With the growing strength of the people opposing the project, the ruling political party began suffering from fear of losing political authority on the area.

7

Summary and Conclusion

In the foregoing presentation various aspects of the protest and agitation over the issue of Nandigram land acquisition have been described and discussed in the context of socio-cultural life, economy of the people under the influence of various geo-historical backgrounds with the contemporary changes in the political situation of the area.

(a) The Important Emerging Agro-economic Issues

1. Benefits of Operation *Barga* have been diluted much and lost viability in most cases in Nandigram.
2. Large number of landless and marginal holders does not have economic and food security.
3. High yielding verities of crops and vegetables have replaced traditional crops, e.g., local paddy and jute with a considerable impact.
4. Unirrigated potato and *Khesari* (pulses) has increased the returns from land recently turning single cropped *Jaljami* (low agricultural land) into fully double cropped.
5. Introduction of brackish water fishery since last two decades has opened up a huge economic opportunity utilising the so far poorest land and saline marshy land.
6. Situation driven illegal land transactions and tenancy *(chuktichas)* have become rampant causing socio-political tension.

From the above situation and analysis of the agro-economic conditions, this area cannot be taken as agriculturally backward.

Instead, the present trends of extended cropping seasons and cropping intensity as discussed above are enough as indicators of a growing agro-economy eventually enhancing the carrying capacity of the land. Hence, the currently available official record of land resource seems outdated and a comprehensive land use and economic survey is urgently needed for assessing the real value of the land at present. On the other hand, updating of the land use database and minimising the illegal farming practices will generate additional revenue and may be helpful to mitigate the complicacies in the land tenure scenario.

(b) The Issues Emerged from the Socio-cultural, Political and Religious Dimension

1. Majority of the villagers are not in a position of getting proper compensation. Rather, they would be evicted without any kind of compensation or a small amount. This kind of anxiety is due to survival threat.
2. The villagers were not ready to give up the security of subsistence and invite a crucially threatened life.
3. The owners of betel leaf cultivators were opposing the land acquisition because of loss of cash crop of this area.
4. The contractors of the fisheries were conspicuous due to good investment and uncertain return.
5. Leadership was developed within the mass. The faith on leadership in both side was intact and consolidated which culminated in triggering wide spread violence and fierce fighting.
6. Religious and cultural affinities were used as symbolic gesture in whole movement to maintain solidarity cutting across the ethnic dimension.
7. It is observed that during this movement the village society was divided into two sections: (a) one section showed their loyalty towards the then ruling party; and (b) the other section of the society, irrespective of their caste, class and religion collectively resisted the proposed land acquisition under the banner BUPC.
8. Many villagers expressed that they were the supporters of the then ruling party but they do not want to give their land

for any industry owing to the failure of Jellinghum project in their locality.

9. The case studies show that affected families have suffered a lot almost in every sphere, i.e., social humiliation, mental unrest, physical assault, economic disorganisation and political upheaval.
10. The villagers have gradually overcome the situation, adjusted with the local village society. However, such types of incidents have created a deep rooted impact on the villagers' minds.

(c) Issues Related to Health and Nutrition

From the age-sex structure of the population a relative threat of low frequency of pre-reproductive age group (26.59 per cent) is observed.

1. A considerable reduction of male members in population structure is noticed particularly in the age group 25-29 years, the most economically and reproductive active population which may be explained due to out- migration or proportionally high mortality.
2. Unmarried male (41.99 per cent) is considerably higher than that of female counterpart (34.65 per cent) which indicates that they are not getting suitable brides from their neighbouring areas.
3. The widowhood and separation in women is significantly higher (5.82 per cent) than the male counterpart (0.44 per cent), this may be explained either due to relatively high mortality of male partners or out- migration (both voluntarily and forced situation) from this area.
4. With regard to the nutritional status, among the children below five years 17.19 per cent were in acute malnutrition and 2.01 per cent in chronic malnutrition which demands an urgent medical intervention.

(d) Psycho-social Stress and Strain

1. The anxiety and depression were the main psycho-social stress developed within the people in Nandigram area at that time.

2. Constant fear psychosis disturbed the villagers in maintaining normal life at that time.
3. The result of such psycho-social stress was the disorder in psyche and physique by manifestations, viz., disturbed sleep, stomach disorder and loss of appetite etc.

In spite of having such psycho-social impact, the people of Nandigram were found to be still in mood of actively participating in such protest knowing fully the outcome.

Who and what caused the Nandigram violence may not be conclusively proved for years. But there are essentially two separate contrasting interest groups having widely varied view points from one another. Each has its merits and lacunae as mentioned in the course of this report. The first scenario is the accounts of intense spontaneous protest and agitation over the issue of proposed land acquisition under Land Eviction Resistance Committee and the other is the top leadership of the then ruling party, determined to rule out all oppositions against industrialisation. However, the villagers of Nandigram still believe that such a tragic violence could have been avoided with a mutual trust and proper mitigation plan.

The movement itself is an account of the peoples' reaction to such acquisition drive irrespective of location and community involved in our country. Hence, the present report, instead of its limited scope, seems to extend a documented evidence of an important instance of such movement in the perspectives of their demographic history, social, economic and political dimensions. The future development planners and executors may find it handy while undertaking any such land acquisition programme in future.

Notes

1. In 16th century, Nandigopal Chowdhuri was the landlord of Gumgarh Parganas. By his name, the name of the village was Nandigram.
2. Professor P.K. Bhowmick defined the frontier tract as a boundary or buffer region, which generally witness many migrations, aggressions and conquests; p. 4; in '*Socio-Cultural Frontier Bengal* by P.K. Bhowmick; Punthi Pustak, 1976.

3. The word *Harmad* is originated from Portuguese word *Armada* that means a kind of pirates who were engaged in looting, ransack, rape and conversion into Christianity; p. 95; in article titled "Hijilite Harmad" by Dr Premananda Pradhan in the book *"Khejurir Sekal Akal*" (Bengali) edited by Swapan Kumar Mondal; Netaji Pathchakra, Purbachura: P.O. Tikasi, Purba Midnapur, December, 2009.

References

Bhowmick, P. K., 1976. *Socio-Cultural Frontier Bengal.* Kolkata: Punthi Pustak.

Bhunia, Pijush Kanti (ed) 2010. *Nandigramer Mati-O-Manush.* (in Bengali).

Bose, Kaushik, Bhunia D., Paul G. *et al.*, 2006. Age and sex variations in undernutrition of rural Bengal primary school children of East Midnapore District, West Bengal, India. In: Ecology, Culture, Nutrition, Health and Disease. Kaushik Bose (ed). *Human Ecology Special Issue*, 14:71-75.

Camp Sarate Nirdesh Danni, Jeray Dabi Praktan DG (in Bengali) *Anandabazar Patrika* May 25th, 2012, p. 7.

Census of India 2001. Primary Census Abstract, Medinipur

Cernea, Michael 1999. Why Economic Analysis is Essential to Resettlement: A Sociologist's View. In: *The Economics of Involuntary Resettlement: Questions and Challenges*. Michael Cernea (ed). Washington DC: The World Bank. pp. 5-49.

Chakraborti, Anil K., Apurba Kumar Mukhopadhyay and Debesh Roy 2003. *Beneficiaries of land Reforms: The West Bengal Scenario*. Kalyani, Nadia, West Bengal, and Spandan, Kolkata: State Institute of Panchayats and Rural Development.

Chakraborti, S. B., 1985. Peasants' environment and Planners' enigma: a paradigmatic approach. In: *Agrarian Situation in India, Vol. II*. Chakraborti, S.B., Ghosh, B.R., Danda, AK (eds). Calcutta: Anthropological Survey of India.

Das, R. K., 2011. Anthropology of Governance. *Journal of the Anthropological Survey of India*, 46, p. 127.

Goswami, B. B. 1979. *The Mizo Unrest – A study of politicisation of Culture*. Jaypur: Aalekh Publishers.

Guha, Abhijit 2007. *Land, Law and the Left: The Saga of Disempowerment of the Peasantry in the Era of Globalisation*. New Delhi: Concept Publishing Co.

Long Live People's Resistance in Nandigram. *Manthan Samayiki, (a bi-monthly Bengali little magazine from Kolkata, West Bengal)*, Thursday, April 5, 2007.

Maity, Tapas 2006. *Medinipurer Gramer Katha* (in Bengali). Madinipur: Upatyaka.

Nandi, J., Sarkar Shamik, *et al.*, 2007. Long Live People's Resistance in Nandigram. *Samikhak Samanway*, Thursday, April 5, 2007.

Nandigram: What Really Happened? Based on the Report of the People's Tribunal on Nandigram on 26-27 May 2007. Daanish Books, Delhi.

O'Malley L.S.S. 1913. *Bengal District Gazetters, Medinipur, 1911*. Calcutta.

Pradhan, Premananda 2009. "Hijlite Harmad" in *'Khejurir Sekal Ekal'* (in Bengali); Swapan Kumar (Ed.), Purba Medinipur: Netaji Pathchakra.

Price J.C. 1874. *The Chuhar Rebellion of 1799* from District Census Hand Books, Medinipur, (Census 1951).

Sarkar, B.N., Prabir Chodhuri, Subhra Bhattacharya, Tapas Biswas *et al.*, 2012. Prevalence of Beta Thalassaemia and other haemoglobinopathies among the adolescent boys and girls of Nandigram, Purba Medinipur. (Unpublished).

World Health Organisation Multicenter Growth Reference Study Group 2006. WHO Child growth standards. Length/Height-for-age, weight-for-length, weight-for-height and body mass index-for-age- methods & development. Geneva: World Health Organisation.

PHOTOGRAPHS

Fig. 1. Village Sonachura at the time of harvesting aman paddy.

Fig. 2. Village scenario depicting the sweet water pond, betel vine (far left), coconut plantation and cultivable land of Sonachura Jalpai.

Fig. 3. The memorial structure for the departed souls of the Nandigram Movement at the Sonachura village market.

Fig. 4. Talpati Khal (tidal creek), an important waterway, dividing the Blocks Nandigram I and Khejuri I & II.

Fig. 5 Typical brackish water prawn bheri with the lucrative catch (inset) of Sonatura Jalpai village.

Fig. 6. Remnants of Jellinghum project by the side of Hugli River at Nakchira Char village.

Fig. 7. Focus group discussion with the villagers at the Kalicharanpur Panchayat office.

Fig. 8. An experienced Mahishya fisherman knitting his net at Sonachura Jalpai village.

Fig. 9. Tailoring as an economic pursuit.

Fig. 10. Children brought by their mothers at the health survey camp at Garchakraberya village.

Fig. 11. Recovered from trauma the joyful children queued up for health check up at the Garchakraberia camp organised by the study team.

Fig. 12. An affected family regaining normalcy from loss of some male members in the violence.

Fig. 13. Emerging un-irrigated potato farming at the harvested aman paddy field. The potato seedling are covered under straw to retain the remaining moisture.

Fig. 14. The village level health workers playing an important role in restoring health in the affected villages.

Fig. 15. Interviewing an affected family during field investigation at Sonachura village.

Fig. 16. The study team from Anthropological Survey of India.

Index

C

D

E

F

G

H

I

J

K

L